This book is to be returned on or before
the last date stamped below.

Planning Pollution Prevention

Planning Pollution Prevention

A comparison of siting controls over air
pollution sources in Great Britain
and the U S A

Christopher Wood

Senior Lecturer in Planning and Landscape
The University of Manchester

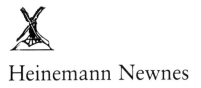

Heinemann Newnes

Heinemann Newnes
An imprint of Heinemann Professional Publishing Ltd
Halley Court, Jordan Hill, Oxford OX2 8EJ

OXFORD LONDON MELBOURNE AUCKLAND SINGAPORE
IBADAN NAIROBI GABORONE KINGSTON

First published 1989

© Christopher Wood 1989

British Library Cataloguing in Publication Data
Wood, Christopher, *1944*—
   Planning pollution prevention: a comparison
   of siting controls over air pollution sources
   in Great Britain and the U.S.A.
   1. Atmosphere. Pollution. Control.
   Role in town planning
   363.7′392

ISBN 0 434 92274 9

Printed and bound in Great Britain by
Redwood Burn Limited, Trowbridge, Wiltshire.

In memory of
Cyril Wood
and for
Norah Wood

Abbreviations

BACT	Best available control technology
BPM	Best practicable means
CEC	Commission of the European Communities
CEQ	Council on Environmental Quality
DOE	Department of the Environment
EIA	Environmental impact assessment
EIS	Environmental impact statement
EPA	Environmental Protection Agency
FONSI	Finding of no significant impact
HMIP	Her Majesty's Inspectorate of Pollution
HSE	Health and Safety Executive
LAER	Lowest achievable emission rate
NEPA	National Environmental Policy Act
NIMBY	Not in my back yard
NSPS	New source performance standard
PSD	Prevention of significant deterioration
RCEP	Royal Commission on Environmental Pollution

Contents

Preface	xi
1 INTRODUCTION	**1**
• Pollution: prevention or cure?	1
The pollution source siting process	3
• Problems in preventing pollution	5
Comparing preventive pollution controls	7
Evaluating US/UK siting controls over air pollution sources	11
• Improving the planning of pollution prevention	13
Notes	14
2 PREVENTIVE CONTROLS OVER AIR POLLUTION	**16**
Air pollutants and their effects	16
Air pollution standards	21
Environmental impact assessment and air pollution control	24
Technical controls over air pollution	25
Land use controls over air pollution	26
Notes	31
3 PREVENTIVE CONTROLS OVER AIR POLLUTION IN THE UNITED STATES	**32**
Context	32
Air pollution control	41
Land use control	52
Environmental impact assessment	63
Notes	66
4 IMPLEMENTATION OF PREVENTIVE CONTROLS IN THE UNITED STATES	**68**
Relationship between land use and air pollution controls	68
The Louisiana creosote storage facility	71
The Texas asphalt plant	74
The North Carolina oil refinery	76
The Florida resources recovery facility	78

viii *Planning Pollution Prevention*

The Maryland solvent recycling plant	81
The Oregon energy recovery facility	83
The California refinery modification	85
The California chemical production facility	87
Notes	90

5 PREVENTIVE CONTROLS OVER AIR POLLUTION IN THE UNITED KINGDOM — 91

Context	91
Air pollution control	95
Land use planning control	108
Environmental impact assessment	114
Notes	116

6 IMPLEMENTATION OF PREVENTIVE CONTROLS IN THE UNITED KINGDOM — 118

Relationship between land use and air pollution controls	118
The new town glass fibre works	123
The Tameside resin manufacturing mill	125
The Bolton sewage sludge incinerator	126
The Yorkshire chemical formulation plant	128
The Bolton lead battery plant	130
The Cheshire fertiliser plant extension	133
The St. Helens sulphuric acid works	135
The Glossop molybdenum smelter	137
Notes	139

7 OUTCOME OF THE SITING PROCESS FOR AN AIR POLLUTION SOURCE — 140

The developer	143
The air pollution control agency	148
The land use planning agency	153
The objectors	160
The authorisation process	166
Implementation of controls	167
Notes	171

8 COMPARING US AND UK PREVENTIVE POLLUTION CONTROLS — 173

The developer	173
The air pollution control agency	179
The land use planning agency	183
The objectors	191
The authorisation process	196
Implementation of controls	198
Notes	201

9 CONCLUSIONS 202
Evaluating US/UK siting controls over air pollution sources 202
- Identifying relative shortcomings in the US and UK
 preventive control systems 210
Improving the planning of pollution prevention 213
Notes 223

Bibliography 224

Index 236

Preface

The idea of undertaking this study grew from previous work on planning and pollution in the United Kingdom. The concepts of ambient pollution standards, increased public participation, freedom of information, environmental impact assessment, etc, had been put forward as means of improving British pollution prevention and it seemed useful to investigate how they affected the siting process for new pollution sources in the country from which many of them originated: the United States of America. As a result of a comparative investigation, it was hoped that relevant recommendations for improving practice in the anticipatory control of pollution in both countries might be advanced.

It was thus necessary to become acquainted with the legal and administrative provisions relating to land use planning and pollution control in the USA and to undertake a number of detailed case studies to determine how these provisions were implemented. I gratefully acknowledge the support of the Economic and Social Research Council which awarded a personal research grant to spend a year in the United States. Additional support was provided by the Sir Herbert Manzoni Trust, the National Society for Clean Air and the University of Manchester, to which bodies gratitude is also due. In addition, I thank the land use planning departments at George Washington University, the University of North Carolina at Chapel Hill, the University of Miami, the University of New Orleans, the University of Texas at Austin, Portland State University, and the University of California at Berkeley for providing the necessary facilities and – frequently – stimulation, information and help.

I also wish to record my thanks to the many people who have offered their encouragement and advice. While it is invidious to single out only a few individuals I am compelled to acknowledge the help of Norman Lee (Economics Department, University of Manchester), Tim O'Riordan (School of Environmental Sciences, University of East Anglia), Judith Rees (Geography Department, London School of Economics), and, particularly, David Robinson (Department of Planning and Landscape, University of Manchester) in Britain. In the United States, David Callies of the University of Hawaii, Chris Duerkson of the City of Denver, Mike Enders of the US Agency for International Development, Bob Healy of Duke University, Rich Liroff of the Conservation Foundation, David Vogel and Mel Webber, both of the University of California at Berkeley, were especially helpful. To the many other individuals who gave me their time and patience, read and commented on drafts of case studies, or helped in other ways, I record my most sincere thanks. Much of the strain of completing this book has fallen on Jo Wood, without whose support I would not have been able to undertake it, and on Jackie Jollie and Anita Tomlinson without whom it would not have been typed.

Christopher Wood
Manchester
August 1988

1
Introduction

POLLUTION: PREVENTION OR CURE?

Pollution[1] is an inevitable consequence of most human activity. The Commission of the European Communities (CEC, 1979) has put this succinctly:

> Almost all human activities make some impact on the natural environment, and almost all industrial processes which transform natural resources into products for man's use give rise to some pollution. Acceptance of the reality of this situation is now general, although there are still some who call for a removal of all pollution, not realising that this would signal the end of human activity, as well as of industrial civilisation as we know it (p 49).

Despite the popular belief that pollution is getting worse, the available information indicates that trends vary greatly between individual pollutants. In general, acute local pollution has become very rare, whereas widespread low-level pollution has been increasingly recognised as a potential problem. Localised elevated pollution levels associated with particular industrial and other stationary sources still cause serious (and sometimes avoidable) problems.

It is generally accepted that pollution prevention is better than cure and, in the phrase that Royston (1979) has popularised, that 'pollution prevention pays'[2]. Royston gave many anecdotal examples of this maxim and the European Commission stated that 'several studies show that the cost of preventing pollution and nuisances is less than the cost of repairing the damage caused and introducing anti-pollution measures' (p 49).

The Commission (1979) found that:

> Too much economic activity has taken place in the wrong place, using environmentally unsuitable technologies. The consequence has often been a choice between accepting pollution as a necessary evil or paying very large sums for its elimination (p 49).

The Commission's environmental policy is overtly directed towards prevention by anticipatory, or prospective, control: 'the best environmental policy consists in preventing the creation of pollution or nuisances at source rather than subsequently trying to counteract their effects' (CEC, 1977:6). This theme of anticipatory action is now widely recognised and is being promoted by a number of international bodies, such as the International Union for Conservation of Nature and Natural Resources (1980). Planning pollution prevention ie, incorporating pollution controls at the planning stage during the siting process for a pollution source, is one of the cornerstones of sustainable development.

While it has not formally endorsed the policy of anticipatory action to forestall pollution, the British government was a signatory to the European environmental programme and generally supports the principle that prevention is better than cure. In the United States of America, also, this concept has wide currency. For example, the central purpose of the National Environmental Policy Act 1969[3] is to ascertain the environmental damage likely to be caused by a federal action before it is taken, in order that it may be modified, abandoned or proceeded with in the full knowledge of the consequences.

Needless to say, the intention to prevent does not invariably preclude the necessity to cure. However carefully considered, prospective controls cannot always anticipate either changes in technology or future trends in production, which may result in unexpected pollution levels, or changes in public attitudes, which may lead to the decreasing acceptability of once-tolerated levels. Similarly, it must be remembered that achieving pollution control compliance in the first instance is no guarantee that it will continue indefinitely. Consequently most countries adopt a two-pronged, mutually reinforcing, approach to pollution control. Holdgate (1979) has summarised this as being:

1 through a land use planning or development control process in which the distribution of sources of pollution is adjusted so as to be compatible with other priority land uses, and so that pollution from new development is constrained from the outset;
2 through controls, operated by various official agencies or voluntarily within industries, limiting existing sources of pollution and ensuring that new sources comply with conditions imposed when they are built (p 202).

Train (1976) stated that the fundamental process to be followed, notwithstanding the comprehensive nature of air pollution controls in the USA, was a combination of emission limitations and land use planning. Morell (1974) has expressed this two-pronged approach succinctly:

> A judicious combination of pollution control technology and more responsible land use decision making in the Environmental Age provides the only effective, long term solution to the problem of air pollution (p 19).

Land use planning controls are primarily preventive (prospective or anticipatory) whereas technical and other (eg, housekeeping) source controls

Introduction 3

(Holdgate's second category) can be both preventive and retrospective (curative). Both are brought to bear during the siting process.

THE POLLUTION SOURCE SITING PROCESS

The following outline of the siting process for new or modified pollution sources is essentially a simplified account of the process through which a developer must progress in order to gain the permits necessary to construct and operate a new source of air pollution. This enables the principal actors in the permitting process to be identified and demonstrates where the various available anticipatory controls can be applied and where difficulties are likely to arise.

Several permits are usually required before construction and operation of a new or modified stationary source of air pollution can commence. The siting approval process has been categorised by Ducsik (1983) as 'decide-announce-defend'. Basically, the *developer* (who may represent either a private company or a public body) chooses a site on the basis of engineering, technical, financial, legal and (sometimes) environmental considerations. He then often proceeds, usually secretly, to acquire the land or an option to purchase it. There may then follow a period of informal negotiation with several agencies, including the *air pollution control agency* (which may be local, regional and/or national) and the *land use planning agency* (which is usually, but not always, local). The developer then announces his project, frequently as an immutable proposal, and applies for the necessary permits from the control agencies (Figure 1).

He will often propose prospective pollution controls but there is normally a period of negotiation between the developer and each of the agencies, during which the developer may be asked to make modifications and the agency may decide to relax some of its requirements, where it has the freedom to do so. As the various environmental and other impacts – including air pollution – become widely known, the public, which has no reason to expect any flexibility in the developer's position, often sets about trying to delay or stop the project by all the legitimate means available to it, including subjecting the control agencies to pressure to refuse the relevant permits. These *objectors* may, for example, be local residents, local environmental groups and/or national environmental groups. There may often also be supporters of the project[4]. On the other hand, some applications may arouse no controversy at all.

The next stage in the authorisation process is an agency decision which, if it is an approval, will almost always be accompanied by conditions which may make further modifications to the proposal necessary. The air pollution control agency's conditions will, of course, include prior controls over pollution, often utilising both equipment requirements and emission standards.

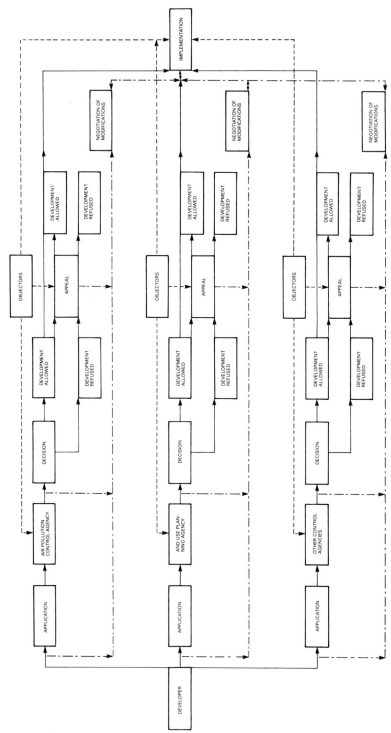

Figure 1 The siting process for a stationary source of air pollution

The land use planning agency's permit may also often be accompanied by similar anticipatory controls over air pollution. At any event, this agency's decision will affect air pollution levels and the consequent damage (Chapter 2).

If the developer or (in some countries) the objectors are unhappy with the decision there may well be an administrative or legal appeal against it during which further negotiations may take place and following which additional controls and modifications may be required. Finally, once all the permits have been obtained, the developer may proceed to construct the facility though there may be further protest from the objectors both during construction and during subsequent operation, especially if air pollution problems occur. Yet more negotiation between the developer and the control agencies may be necessary in the process of implementing their conditions.

PROBLEMS IN PREVENTING POLLUTION

The main problem of utilising preventive controls over pollution is that they do not always work satisfactorily. Apart from the difficulties associated with changing technology, production levels and public attitudes, the preventive controls applied may fail to mitigate pollution levels adequately and serious damage may occur. While such damage may sometimes result from teething problems in the commissioning of new plant, it is often more deep-seated. It may well be caused by a failure to anticipate both the likely pollution concentrations following the application of controls and the effect of that pollution in the particular locality concerned.

In the United Kingdom, for example, both the use and the effectiveness of land use planning controls over pollution have been somewhat variable. The Royal Commission on Environmental Pollution (RCEP, 1976), while recognising that many considerations other than pollution were important in making planning decisions, was very critical: 'Our concern is not that pollution is not always given top priority; it is that it is often dealt with inadequately, and sometimes forgotten altogether in the planning process' (p 93). The conclusion that planning practice in the control of pollution frequently leaves much to be desired has been borne out in a number of studies (eg Wood, 1976; Wood and Pendleton, 1979; Ledger, 1982; Miller and Wood, 1983). While there is some evidence that the attention paid to potential pollution problems by local planning authorities may be increasing, there is still ample scope for improvement (Miller and Wood, 1983).

In the United States there have also been many inappropriately sited polluting industrial establishments and many instances of sensitive receptors being located too close to existing sources of pollution (Arnold and Edgerley, 1967). These mistakes have occurred in localities with land use planning controls as well as in jurisdictions without them.

6 *Planning Pollution Prevention*

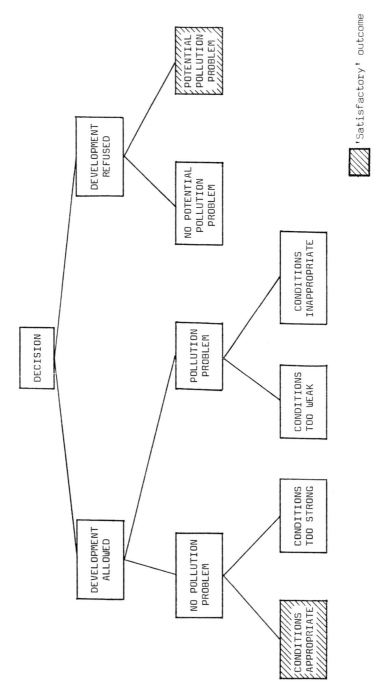

Figure 2 Outcomes of the use of controls over pollution during the siting process

If there have been numerous instances where anticipatory controls over pollution have proved inadequate, there have been others where development which could have proved environmentally acceptable has been prevented, ostensibly because the anticipated pollution concentrations were too high. This is another aspect of the problem of siting new sources of pollution. There have been examples of such decisions in the United Kingdom (Miller and Wood, 1983) but it is in the United States that a tangled skein of environmental legislation has allowed much new development to be stopped by determined opponents, frequently by recourse to the courts, and sometimes with little regard to the actual environmental merits of the issue. Indeed, the term *preventive control* frequently acquires an ironic meaning as development, rather than just pollution, is prevented.

A further aspect of the use of anticipatory controls over pollution is the appropriate selection of conditions appended to any permission. If development is permitted, it could be that no pollution problems arise but the conditions may prove unduly and unnecessarily onerous for the developer. If a pollution problem does occur it may either be because the conditions are too weak or because the development is fundamentally unsuited to the location and *no* conditions would be appropriate. Of the six outcomes shown in Figure 2, only two (the use of appropriate conditions and the refusal of a development which would have caused a pollution problem had it been allowed) could be termed satisfactory decisions. Even the former acceptable outcome may be subsequently marred by permitting the encroachment of sensitive land uses such as residential or educational properties upon the existing pollution source, as well as by the possible temporal changes mentioned above.

There may be a very large number of reasons for failing to reach a satisfactory decision. To take only a few examples, the legal powers available to impose conditions may be inadequate, the enforcement of conditions may prove to be weak, the potential pollution problem may not be recognised and either the developer or the objectors may prove to be unduly influential. It is clear, then, that the appropriate utilisation of preventive pollution control requires not only an adequate armoury of legal anticipatory powers but the conscientious implementation of these powers.

The principal purpose of this book is to suggest ways of overcoming the problems associated with the use of preventive controls over pollution in the siting process and hence to assist in improving the planning of pollution prevention.

COMPARING PREVENTIVE POLLUTION CONTROLS

Perhaps the most effective way of generating suggestions for improvements in the use of prospective controls is by undertaking a comparison of the legal and administrative powers available for preventive control and of the implementation of these powers. As Lundquist (1978) has stated, comparative studies of national approaches to solving environmental problems have often led to valuable and practical suggestions to improve the effectiveness of the national processes examined.

The twin approaches of using land use planning powers and specific pollution control legislation to abate pollution applies in the United Kingdom as elsewhere. Britain has a strong and fairly comprehensive land use planning system and the official position has been expressed by the Department of the Environment (DOE, 1981):

> Planning permission should be refused only where the authority is satisfied that the development in question would result in a significant deterioration of local air quality even after the use of specific powers to control pollution (para 30).

As will be shown in Chapter 5, these specific powers are largely confined to plant by plant controls over certain (but not all) new sources of pollution. Despite official British acceptance of the role of land use planning in the control of pollution, the use of both technical and land use planning controls to implement preventive pollution abatement leaves much to be desired. At present, land use planning decisions affecting pollution levels, like most other British planning decisions, are made on an ad hoc basis by considering the merits of the particular case in question, bearing in mind any relevant national and local policies.

This approach is in marked contrast to that adopted in the United States where there are much stronger prior technical controls over emissions from virtually all new sources of pollution. As will be shown in Chapter 3, the national source-by-source control requirements linked to the use of environmental quality standards to control the location of pollution emitters provides a formidable array of technical powers. Indeed, it is inevitable that the national pollution control legislation will itself encourage land use patterns which will reduce pollution levels (eg, Council on Environmental Quality (CEQ), 1974:31–34; Train, 1976). The use of environmental impact assessment in certain circumstances and the generation and release of detailed information relating to air pollution control proposals encourage the prior evaluation of pollution impacts, which should lead to improvement in the quality of decisions. However, the American system is very complicated and formalised and concern has been expressed by economists and others that the United States 'should abolish reliance on symbolic standards and instead concentrate on pragmatic ways of getting the job done' (Friedlaender, 1978:11).

By contrast with the national air pollution control system, the US land use planning system has never received national endorsement (Lyday, 1976) and is essentially local and disparate in character. Despite the enormous variations in local land use planning controls, and in their implementation, the value of using land use planning legislation to reduce pollution has been recognised for many years (Hagevik et al, 1974:85–89).

In summary, therefore, the nationally homogeneous UK system of preventive controls involves a relatively weak set of formal pollution control powers (with some new industrial sources escaping prior technical control altogether) and a relatively strong set of land use planning powers. The US preventive system, on the other hand, involves strong and reasonably homogeneous national prior pollution control powers but variable and frequently weak land use planning controls. Despite the very substantial governmental differences between the countries, an analytical comparison of the operation of anticipatory controls over pollution in the United States and the United Kingdom should provide valuable insights into the nature of both systems of control and informed recommendations for overcoming some of the acknowledged shortcomings of the preventive control systems in both countries.

There are several advantages in limiting a comparison to air pollution and to the control of stationary rather than transportation sources of air pollution. It is in air pollution control that the archetypal British means of control over industrial pollution, the use of *best practicable means*, is most widely established, and perhaps it is here that the rigid American standard-setting approach is best exemplified. Air pollution is probably the most widely understood and most widespread form of pollution and consequently has most literature devoted to it. This is certainly true of the writing dealing with land use planning controls over pollution in the United States (eg, Kurtzweg, 1973; Kaiser et al, 1974). Finally, most of the existing UK case studies of planning and pollution tend to relate more to air pollution than to other forms of pollution.

There are obvious dangers in studying one form of pollution in isolation. Action taken to reduce air pollution may well lead to increases in other types of pollution, since the wastes from which pollution stems must be disposed of elsewhere (perhaps on land or in water bodies). This multi-media nature of the pollution problem is well exemplified by sulphur dioxide emissions. Low stacks lead to high local air pollutant levels, tall stacks may contribute to acid rain and water pollution, while desulphurisation leads to either solid waste or liquid waste disposal problems which may themselves lead to water pollution. The problems of the indestructability of matter and of the appropriateness of the medium to which waste is discharged must therefore be borne in mind in analysing air pollution control. In the United Kingdom, these problems have received official recognition in the acceptance of the concept of the 'best practicable environmental option'[5]. One of the objectives of the environmental impact assessment system in the United States and elsewhere has been to endeavour to contribute to overcoming these problems (Wood, 1988).

The same difficulty in isolating air pollution as the subject of study applies to the objectives of the control agencies. It has to be accepted that those concerned with administering air pollution control powers frequently cannot implement these without regard to other objectives. Thus, air pollution controllers are normally only too aware that local employment may hinge on their decisions and they therefore interpret their regulations (to the degree that they are flexible) to permit industrial enterprises to operate but to control pollution from them.

Land use planning is, by definition, a multiple-objective activity (Hall, 1983). The control of air pollution, and indeed of all types of pollution, is merely one desirable aim in a whole array of environmental, social and economic land use planning goals. Thus, air pollution control, while laudable, will seldom be the sole objective of a planning decision and may be neglected because other objectives (such as job creation, visual amenity, etc) are given more weight. Having said this, it should be the case that air pollution control is explicitly considered where pollution is likely to be a problem, even if it is subsequently outweighed by other land use planning goals. At any event, decisions in which air pollution is an important issue are not uncommonly locally controversial and have frequently dominated political agendas (Blowers, 1982). This multiplicity of objectives of the controlling agencies renders quantitative comparative analysis virtually impossible because of the number of factors involved in making the relevant decisions and leads inexorably to the use of case studies to provide the basis for comparison.

Masser (1986) has eloquently justified the use of case studies in comparative studies. Certainly, the case history has become a well-established method of studying developments with environmental repercussions in both the United Kingdom (eg Gregory, 1971; Kimber and Richardson, 1974; Smith, 1975; Miller and Wood, 1983; Blowers, 1984) and the United States (eg Crenson, 1971; Caldwell et al, 1976; Duerkson, 1982). Hall (1982) has undertaken both British and American case studies of planning decisions.

There have been several Anglo-American comparisons of different aspects of environmental control. Clawson and Hall (1973) undertook a major comparison of planning and urban growth. Haar compared the legal background to land use control in the two countries in 1964 and returned to the theme more recently (1984). Garner and Callies (1972) have also explored this area thoroughly. The most recent and relevant comparison of environmental regulation in the US and the UK is that by Vogel (1986).

Other comparative environmental control studies involving Britain or the United States of America, or both, include those by Rose (1974) on urban change, Enloe (1975) on pollution control, Corden (1977) on new towns, Lundquist (1980) on air pollution, Kelman (1981), Kunreuther and Linnerooth (1983) on siting gas facilities, Downing and Hanf (1983a) on air pollution and Peacock (1984), Badaracco (1985), Brickman et al (1985), Wilson (1985) and Masser and Williams (1986) on urban planning.

The main difficulties in undertaking case studies are lack of representative-

Introduction   11

ness and inaccuracy. The problem of unrepresentativeness can never be completely overcome. The case histories recounted in this book, which all concern specific development involving new industrial or public sector sources of air pollution which were proposed but which may or may not have been constructed, were to some extent chosen because of their intrinsic interest rather than to be representative of the activities of the authorities concerned. To reduce this problem, the case studies must be set firmly in their context. In undertaking this comparison, the literature was reviewed, interviews were carried out with officials at both local and higher levels of government, with researchers in universities and research institutes, with industrialists and with pressure group campaigners, apart from those undertaken in connection with particular case histories.

In compiling the case histories, an attempt was made to overcome inaccuracies by obtaining accounts from more than one protaganist in every instance, by checking documentary evidence and by sending longer drafts of the case studies to several participants so that factual details could be verified[6].

EVALUATING US/UK SITING CONTROLS OVER AIR POLLUTION SOURCES

There is always a danger that the excitement of discovery will deter the case study author from making comparisons and drawing conclusions. The case studies are merely a means to an end: the comparison of American and British siting controls over air pollution. That comparison is, at least partly, aimed at answering the question 'do they do it better?' To answer that question an evaluation needs to be undertaken of the two systems.

As Mitchell (1979) has stated:

> Whenever a value judgement is made about a policy, programme or project, such decisions are based on criteria, whether implicit or explicit ... measuring or operationalising the criteria represents a major obstacle in evaluation research ... Ideally, a variety of criteria should be used when judging the adequacy of a given programme (p 258).

While there are numerous criteria which could be adopted for judging the national systems of anticipatory air pollution control, including flexibility, responsiveness and comprehensiveness, those mainly utilised here are the classical measures of efficiency, equity and effectiveness. There is, in practice, some overlap between the requirements of these three criteria and conflicts can frequently arise in satisfying each of them (Rees, 1985).

Efficiency has many meanings but here it is used to measure the relationship between the benefits and costs arising from a decision or regulation: it is an indicator of welfare. An allocation of benefits and costs is said to be efficient 'if it is impossible to move to another allocation which would make some people better off and nobody worse off' (Begg et al, 1984:235).

It is a condition of economic efficiency that polluters should bear the costs imposed on second and third parties and incorporate these in output prices. This is the 'polluter-pays principle' (Organisation for European Co-operation and Development, 1975). One way of bearing the residual costs remaining after the use of pollution control technology is to pay compensation to those directly affected by the pollution. Though the use of direct financial payments is often controversial, it can be consistent with the attainment of economic efficiency:

> A company planning to construct a plant ... should be obliged to take into account not only investment costs but also the social costs of production ... If it is possible to compensate for the negative external effects completely, by direct negotiation between those on whom damage is inflicted and those inflicting it, the negotiated solution then acquires the allocative function of an efficient pricing system, whereby only those projects can be implemented which can produce a surplus on returns over social costs (Knödgen, 1983).

Because of the difficulties in establishing the levels of control at which the efficiency criterion is satisfied, some jurisdictions endeavour to minimise the costs of damage, leaving polluters to argue the case against the marginal costs of increased control. They thus choose some level of control, often on the basis of standards (Chapter 2), in order to avoid having to try to calculate the benefits of control in any particular instance. *Cost-effectiveness* – the cost of achieving a given level of control – is then used to give a partial measure of efficiency. This cost includes both the expense of the permit application procedure (to the developer and to society) and the expense of control attributable to the developer. Cost-effectiveness should not be confused with effectiveness.

Given the lack of information about the national costs of control and of air pollution damage, it will seldom be possible to make other than general observations about the efficiency or cost-effectiveness of the two control systems.

There are two distinct aspects of *equity* which must be considered as evaluative criteria: outcome equity and procedural equity. *Outcome equity* is distributional, ie, it is concerned with who bears the costs of, and who benefits from, the outcome in any particular case. It has to do with fairness. *Procedural equity* demands that the decision-making process should be fair and be seen to be fair. In a procedurally equitable system, the developer, the local population and the wider public should have similar rights to participate in the decisions affecting them.

Procedural equity may demand advertisement and other publicity, consultation, the making of representations, public hearings or inquiries, appeal procedures and other appropriate means. While some objectors to a proposal may never be satisfied, the vast majority may accept an outcome if their views have been genuinely considered and are reflected in the decision. Because of the costs involved in consultations, hearings, etc, procedural equity frequently conflicts with efficiency.

The *effectiveness* of a pollution control system is a measure of how well it works in practice. Thus, the effectiveness of a decision refers to the extent to which it is complied with and achieves the desired objectives. Effectiveness is thus closely related to the implementation of policies. This, in turn, subsumes the enforcement of any sanctions against non-compliance.

It will often be necessary to make trade-offs between efficiency and effectiveness. As Downing (1982) stated:

> Economic realities require some balancing of benefits and costs. When this is done during implementation, no matter who is responsible for action, laws which ignore costs will be compromised. Implementation deficits may result from efficient adjustment to economic realities or both. In any case they appear to be inevitable and universal.

Similarly, trade-offs may have to be made between efficiency and equity where they conflict (Keogh, 1985). The conflicting requirements of the three criteria reinforce the need to utilize all of them. As McAllister (1980) has argued, the central purpose of evaluations should be to help decision-making, although the information provided will seldom lead to a clear-cut judgement between complex choices (p 277).

It is precisely the lack of availability of an accurate measure of the net benefits of permissions and the consequent difficulty of reaching clear-cut decisions on the control of stationary sources of air pollution that make the role of politics in the process so crucial. As Gregory (1971) observed, after examining a number of celebrated British environmental controversies:

> Undoubtedly, there are some forms of pollution that really are matters of life and death, or will be in the not too distant future. They, of course, are forces of absolute evil, and no price is too great to ward them off. But it would be nonsense to suggest that all threats to amenity are as serious as this. The conflict between industry and amenity takes many forms, but it is not always a contest between virtue and evil (pp 306–307).

IMPROVING THE PLANNING OF POLLUTION PREVENTION

In order to achieve the principal purpose of this book – making suggestions about how to improve practice in planning pollution prevention – several preliminary steps must be made. The first involves investigating the nature of air pollutants and the land use and other anticipatory control techniques for abating air pollution from new or modified sources (Chapter 2). The next requires acquiring an understanding of the legal and administrative powers available to put these techniques into effect in the USA and the UK. Consequently in Chapter 3 there is a discussion of several national characteristics relevant to an analysis of US stationary source control, of the powers available to the air pollution control agencies and the land use planning agencies and of practice in their utilisation. Chapter 5 contains a parallel treatment of the UK.

The third step demands an investigation of the implementation of anticipatory control powers and techniques in the US and UK. To this end the American case studies are presented in Chapter 4 and the British case studies in Chapter 6 and these provide the basis for completing the remaining steps.

The fourth step is to develop an understanding of the siting process for new air pollution sources in the US and UK. In Chapter 7 the common elements of the British and American systems of pollution prevention are analysed according to the role of the developer, the air pollution control agency, the land use planning agency and the objectors. The siting process as a whole is then discussed. Finally, there is an examination of the factors determining the success with which controls on new sources are implemented. This analysis leads to the derivation of six indicators of the outcome of the siting process for a new or modified source of air pollution which hold true in the British and American cases studied.

The fifth step is to compare and contrast the US and UK anticipatory control systems. In Chapter 8 the roles of the developer, the air pollution control agency, the land use planning agency and the objectors are used to explain the differences between the two systems. The overall authorisation processes, and the implementation of controls in the two countries, are also compared.

The sixth step is the evaluation of American and British siting controls over new sources of air pollution. This is presented in the first part of Chapter 9. The seventh step is to reveal, by comparison, the relative shortcomings of the US and UK preventive pollution control systems. This identification forms the second part of Chapter 9 and leads directly to the making of recommendations to overcome perceived shortcomings. The adoption of these suggestions should lead to the fulfilment of the purpose of this comparative study by assisting in improving the utilisation of anticipatory pollution controls: planning pollution prevention.

NOTES

1 Pollution may be defined as 'the introduction by man of waste matter or surplus energy into the environment, directly or indirectly causing damage to persons other

Introduction 15

than himself, his household or those with whom he has a direct contractual relationship.'
2 Royston claimed that 'adding on' pollution control equipment is about 3 or 4 times more expensive than 'building in' the same controls.
3 *National Environmental Policy Act 1969* 42 USC 4321–4327.
4 Kunreuther and Linnerooth (1983:10) distinguish between local residents, who may be expected to support a project having local benefits, and public interest groups who may be expected to oppose it.
5 This phrase was coined by the Royal Commission on Environmental Pollution (1976:76) which advocated its use in controlling the various forms of pollution from a new source. The Commission devoted its twelfth report (RCEP, 1988) to the topic, where it defined BPEO as follows:

> A BPEO is the outcome of a systematic consultative and decision-making procedure which emphasises the protection and conservation of the environment across land, air and water. The BPEO procedure establishes, for a given set of objectives, the option that provides the most benefit or least damage to the environment as a whole, at acceptable cost, in the long term as well as in the short term (p 5).

6 The only case study not undertaken or supervised by the author was conducted using the same approach and a much longer version than that summarised here has been published (Duerkson, 1982).

2
Preventive Controls Over Air Pollution

Figure 3 makes it clear that pollution is the result of a process originating in the generation of waste. Both technical source controls and land use planning controls can be applied at various stages in the pollution process. In principle, at least, both types of control may be applied both retrospectively (once a pollution problem has been shown to exist) and prospectively (to prevent a pollution problem from arising or to limit its magnitude). In practice, however, only technical and other source controls can usually be used retrospectively (on existing sources) and then perhaps to a more limited extent than might be expected because of the very high costs of installing control devices after commissioning plant, rather than at the outset. Land use planning controls tend, by their very nature, to be anticipatory though examples of their retrospective application do exist.

This chapter is divided into five parts. The first commences with a brief description of various air pollutants and their effects, together with a discussion of the difficulties of quantifying the costs and benefits of control. The next two parts explain the use and relevance of air pollution standards and of environmental impact assessment: both are relevant in pollution prevention or amelioration. The various methods of technical and other source controls available to reduce pollution damage are then briefly mentioned. The fifth part begins by discussing the role of land use planning controls in anticipating pollution problems. Finally, the various land use planning techniques available for controlling air pollution are reviewed.

AIR POLLUTANTS AND THEIR EFFECTS

Wastes are emitted into the atmosphere as either gases or particles, and are eventually removed by natural self-cleansing processes. The wastes mostly originate from the burning of fossil fuels and the processing of materials. Transport (especially the motor vehicle), industry and the domestic sector are all important fuel users and hence sources of pollution (Table 1). In addition, industry emits large quantities of airborne process wastes. Institutional and commercial sources may be significant, particularly in city centres, and

Preventive Controls Over Air Pollution 17

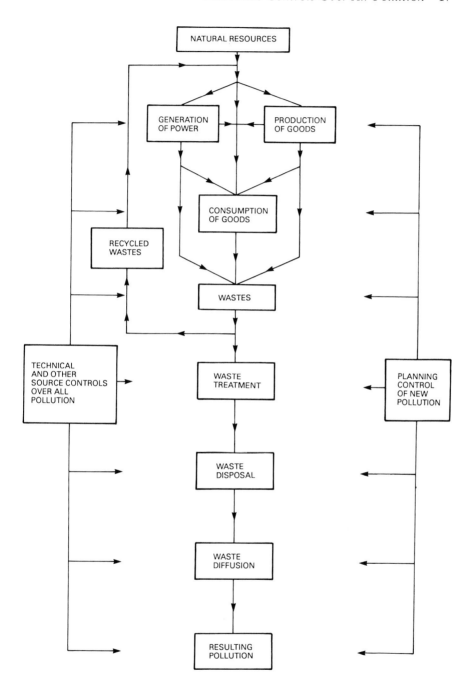

Figure 3 The pollution process and its control

TABLE 1 A summary of air pollution effects

Pollutant	Major sources	Health	Susceptible populations	Vegetation	Materials	Aesthetics/Nuisances	Comments
Carbon monoxide (CO)	Transportation, industrial processes	Reacts with haemaglobin reducing mental attentiveness, physical exertion, and exacerbating cardiovascular disease symptoms	Persons with cardiovascular disease and others	None	None	None	Past knowledge was based on study of high exposure for short periods with healthy, young individuals. New data show possible health effects for susceptible persons at CO levels in the blood found in urban populations
Nitrogen oxides (NO_x)	Transportation, space heating/cooling, power generation	Interfere with respiratory functions producing long-term (chronic) disease symptoms	Persons with respiratory or cardiac disease, the young and the elderly	Reduction in growth of plants with broad leaves (eg beans, tomatoes)	Accelerated deterioration of dyes and paints	Creation of a brownish colouring in urban air	Conclusions are based on limited exposure of healthy adults to low doses, extensive animal studies, and only limited data relevant to ambient conditions
Hydro-carbons (HC)	Transportation & industrial processes	See photo-oxidants	See photo-oxidants	None	None	None	Indirectly polluting through the production of photochemical oxidants upon reaction with NO and NO_2 in the presence of sunlight
Photo-oxidants (O_3)	See nitrogen oxides and hydrocarbons	Interfere with respiratory functions & cause eye irritations	Persons with chronic respiratory diseases, especially bronchial asthma	Severe reduction in growth and eventual death of leafy vegetables, field and forage crops, fruit shrubs, fruit and forest trees caused by ozone and PAN	Ozone causes the cracking of rubber and the accelerated deterioration of nylon, rayon dyes and paints	Ozone has a distinct although not terribly offensive odour	Ozone (O_3) is the most common type and the key indicator for photo-oxidants. Health effects are based on limited and inadequate data. Ozone, peroxyacetylnitrite (PAN), etc, are formed by atmospheric reactions

Pollutant	Sources	Health effects	Effects on vegetation	Effects on materials	Aesthetic effects	Comments
Particulates	Power generation, space heating/cooling, industrial processes, soil erosion	Interference with respiratory functions, possible contribution to lung cancer	Persons with respiratory disease, the young and the elderly	Reduction in plant growth by physical blockage of light when deposited on leaf surface	Soiling of fabrics and buildings and corrosion of metals when combined with SO_2	Creation of smoke plumes, scattering of sunlight to produce haze and colourful sunsets, and formation of hygroscopic nuclei to produce fog	The effects of particulates are difficult to separate from those of sulphur dioxide
Sulphur oxides (SO_x)	Power generation, space heating/cooling, industrial processes	Little effect in the pure gas form; similar effects as particulates when combined with them	Persons with respiratory or cardio-vascular disease, the young and the elderly	Reduction in growth of plants with broad leaves	Corrosion of iron, metals, accelerated deterioration of building stone, cotton, paper, leather, paints and other finishes	Scattering of sunlight to produce haze, production of unpleasant odours	Sulphur dioxide is readily converted to SO_3 and then to sulphuric acid (a particulate). Determining which effects are due solely to SO_2 is difficult in acid rain research
Heavy metals, radioactive agents, carbon dioxide, others	Power generation, industrial processes, space heating	Specific to each pollutant	Specific to each pollutant	Fluoride causes long-term damage to selected field crops (and animals)	Tarnishing of metals by hydrogen sulphide	Hydrogen sulphide produces extremely unpleasant odours	Pollution from these agents can be intense at the source, but tends not to be widespread. However, carbon dioxide is the main cause of global 'greenhouse effect' and chlorofluorocarbons affect the ozone layer

Source: adapted from Keyes (1976)

agriculture may generate high dust concentrations under adverse, windy conditions (Stern, 1977b).

As Figure 3 indicates, the pollutant concentrations to which the emitted wastes give rise (usually measured by the amount of pollutant per unit volume of air at ground level) are determined by the way in which the pollutant is dispersed, one of the main factors being the height at which emission takes place. Gaseous pollutants are generally quickly and uniformly mixed with a large volume of air. The important common gaseous pollutants include oxides of sulphur, nitrogen and carbon, hydrogen sulphide, hydrogen fluoride, hydrogen chloride, hydrocarbons and ozone. Particulate pollutants consist of finely divided liquid or solid matter which may be small enough to remain suspended in the atmosphere for some time. The mainly solid particulates are roughly classified by size as smoke, fume, dust and grit. Their composition varies from unburnt carbon to complex compounds, and they include such toxic elements as lead, beryllium, cadmium and certain radioactive elements, as well as various other substances, for example asbestos (Stern, 1977a).

Air pollutants vary in their effects (Table 1). Air pollution is, however, associated with bronchitis, emphysema and lung cancer, with the corrosion of metals and with the soiling of, and damage to, stonework, painted surfaces and fabrics. Air pollution is also known to affect farm animals and crops adversely, to diminish the number of bird species in a given area, and to reduce visibility. While most of these effects are encountered relatively close to the source of pollution, acid rain (caused by the wet deposition of oxides of sulphur and nitrogen as dilute acids) may occur hundreds of miles from the source, resulting in reduced forestry yields and fishery decline. Photochemical smog, on the other hand, is a regional phenomenon caused by the action of sunlight on hot stable air masses containing oxides of nitrogen and hydrocarbons (Stern, 1977b). Air pollution cannot, therefore, be described as being only a local, regional or supraregional problem. It is frequently all three.

There are considerable difficulties associated with the establishment of damage (dose/response) functions (Saunders, 1976). It is generally extremely hard to determine precisely what degree of damage to a particular receptor is associated with a given concentration of pollution. The accurate identification of the source or sources of a particular pollutant is also very hazardous. Further, the estimation of these effects in terms of measurable parameters (number of days off work, percentage drop in crop yield, etc) is uncertain and the fixing of money values to these parameters involves additional assumptions and controversy. Nevertheless, approximate threshold concentrations (below which most types of damage are minimal) have been established for several air pollutants.

Various attempts have been made to assess the costs of air pollution. These are prone to substantial error and must be treated with great circumspection. The monetary costs of air pollution are, however, known to be very high. In the United States, for example, the range of economic benefits from existing air pollution control based upon improvements to human health, reduced soiling and cleaning costs for households and reduced damage to vegetation, crops and materials was estimated to be in the range of $4,600–51,200M per annum in the late 1970s (National Commission on Air Quality, 1981: 4.1–5).

In principle, it should be less difficult to obtain estimates of expenditure on pollution control than it is to put a monetary value on damage. In practice, however, figures are by no means readily available and those that do exist are prone to substantial error. The figures involved are again large. In the United States in 1984, for example, the total costs related to air pollution control have been officially estimated at $30,000M, of which a substantial proportion was expended on stationary source control[1].

AIR POLLUTION STANDARDS

Ideally, controls over individual sources of pollution should be set at a level where the costs of increasing or decreasing control are both positive. However, while the concept of balancing the costs and benefits of pollution control, and hence determining the type or degree of technical or other controls to be applied, is very attractive (Ridker, 1967), the difficulties inherent in operating a system of control based purely upon an optimum level of pollution from each source have rendered it impractical. Accordingly, control based upon standards, which may be set to attain some empirical objective such as the avoidance of health effects, is more useful and is widely employed. Standards may relate either to air pollution concentrations or to the emissions of wastes and may be determined nationally, regionally or locally. It is possible to classify the various types of standard as:

1 environmental quality standards;
2 source emission standards;
3 area emission standards.

Environmental quality standards refer to limits of ambient environmental quality that cannot be exceeded without infringing statutory law. It is apparent that if an area has pollution levels higher than the environmental quality standard, then no new emissions in that area should be permitted and strenuous action should be taken to reduce existing emissions. In other words, no new sources should be constructed. Such standards therefore have a very profound impact upon land use (Brail, 1975). In practice, a mixture of both stringent anticipatory technical controls and land use planning controls are normally employed to attempt to attain and maintain air quality standards. The use of air quality standards, of course, implies an extensive monitoring system to record ambient concentrations on a regular basis.

Environmental quality standards are widely used in the USA (Chapter 3) and, increasingly, in Europe and the United Kingdom (Chapter 5).

Because of the profound implications of air quality standards for both existing and potential polluters, the use of air quality targets or guidelines has become quite common. These are not legally enforceable and hence cannot be regarded as standards. They are suggested desirable targets at which pollution control authorities should aim and in relation to which improvements in environmental quality can be measured. The World Health Organisation (1987) for example, has issued guidelines for some 20 pollutants. In Britain, the Royal Commission on Environmental Pollution (1976) reported that:

> We have reached the view ... that there is now a need to focus attention openly and specifically on air quality. We do not think that air quality *standards* would be a sensible way of achieving this ... Such a system would not only be impracticable at present, it would also not be justified by current knowledge of the effects of pollutants and the social costs they cause. We propose the establishment of air quality *guidelines* (p 47, emphases in original).

However, there is now a strong European-led movement towards the use of standards.

Source emission standards refer to a numerical limit to the amount of a particular pollutant which may be discharged from a specific source. They may apply to new or to existing sources but are most frequently utilised in their most stringent form for new or reconstructed sources. National emission standards for industrial process or heating emissions will not have any land use implications, but locally determined emission standards obviously may, since they could encourage industry to locate in an area where standards are lowest. Source emission standards normally require some agreed level of technical control.

Source emission standards, which are very widely used, may be determined in terms of concentration of pollutant per unit volume of air, weight of emission per ton of product, weight of emission per ton of raw material, weight of emission per unit of time, percentage removal of emission, etc. The American *new source performance standards* (Chapter 3) are source emission standards, frequently specified in terms of either weight of emission per unit of time or per unit volume of air. The British *presumptive limits* (Chapter 5) are also source emission standards. Emission standards require the measurement of concentrations of pollutants in the effluent gases from a process and this is frequently difficult to achieve accurately.

Environmental quality standards, in theory at least, form the basis of emission standards, since only by limiting or adequately dispersing emissions can concentration targets be met. The relationship between the two types of standard is normally quantified by means of mathematical models which are becoming relatively accessible with modern instrumentation. They are also becoming more accurate, often yielding plume dispersion estimates to within

50% of measured results. In practice, the relationship between the standards is less rigorous, because environmental quality standards may be derived politically, rather than on the basis of scientific analysis of dose-response relationships and risk-benefit considerations (O'Riordan, 1979a; Royal Commission on Environmental Pollution, 1984). Emission standards may then be determined using a general set of objectives which is adjusted in accordance with the best means of control available at a realistic cost.

Area emission standards relate to the total emissions from a given area of land (ie, a collection of sources). One variant is emission density zoning, in which the maximum legal rate of emissions of air pollutants from any given area (perhaps a hectare) is limited by the size of the area (Roberts et al, 1975; Venezia, 1976). The relationship to land use is obvious: new facilities can be evaluated within the context of emissions from existing facilities and benefits can be derived from locational controls and the provision of open space. Another variant is emission allocation in which a maximum legal emission rate is assigned to a large area (often the area controlled by a local jurisdiction) with provisions for sanctions, such as a construction ban, to ensure the area's allocation is not exceeded[2]. Here the precise distribution of pollutants within the overall total is not of crucial concern, but a detailed emission inventory and a clearly formulated land use plan are both essential.

A third variant is the specification of land use, or zoning. By allocating a particular area for residential use or for industrial or commercial use an emission standard is, albeit very crudely, implicitly specified. This concept is refined in the concept of *performance zoning*. Performance standards define the maximum amount of smoke, dust and other pollutants that an industry in a given area may produce. The maximum levels of emissions from such a zone should thus, at least in principle, be accurately known:

> Performance standards for industrial districts should be viewed with caution, however, as some jurisdictions do not have the technical capabilities within the planning or zoning department to enforce the standards ... In other areas the local air pollution control agency can and does enforce performance standards (Voorhees, 1974:19).

Monitoring of emissions will clearly be necessary, as with the use of source emission standards.

It is, then, quite apparent that area emission standards are closely interwoven with land use and with land use planning. The use of technical controls on the sources will also be necessary. There are clearly inequities involved in area emission standards, as in environmental quality standards, in that an existing polluter may be given a continuing licence to emit, whereas a new source, however potentially well controlled, may be prohibited from locating in the area concerned. For this reason area emission standards, with the exception of the implicit specification of emissions by determining the land use for an area, have not been widely employed.

ENVIRONMENTAL IMPACT ASSESSMENT AND AIR POLLUTION CONTROL

The necessity to anticipate potential environmental problems so as to avoid them or to reduce their effects requires the thorough appraisal of an environmentally significant action before it is taken. The formalisation of this concept is embodied in the environmental impact assessment process.

The term *environmental impact assessment* stems from the US National Environmental Policy Act 1969. Because of the generally recognised success of environmental impact assessment (EIA) in the United States (CEQ, 1982a: 171–187), it has provoked widespread interest elsewhere, not least in the Commission of the European Communities, which has approved a directive (1985) on EIA. Basically, EIA involves evaluating the impacts of a proposed development or action on the environment it is likely to affect and the publication of these for widespread review before the decision to proceed with the development is taken. Thus the US system includes provisions for the preparation of a draft and a final environmental impact statement, for consultation, for public participation and for the evaluation of alternatives.

The original intention of the National Environmental Policy Act (at least in principle) was to ensure that all the major impacts of an action significantly affecting the environment were evaluated so that the decision to proceed with, to abort or to modify the action could be taken on the basis of full information. In practice, it is rare for environmental impact statements to be written for projects that are likely to be refused permission so the whole emphasis of EIA has turned to mitigating the impacts of the proposal to make it more environmentally acceptable (Holling, 1978)

EIA normally involves a thorough investigation of both the air pollution and land use impacts of a proposed development (eg, Canter, 1977; Munn, 1983). It can therefore be a useful tool for helping to anticipate the likely effects of air pollution from new industrial sources. While the EIA may indicate either that the project should be approved as proposed or refused out of hand, it is more likely to prove the precursor of both technical and other source controls and of land use controls to allow the development to proceed but to ameliorate its environmental impacts. Mitigation measures may be suggested by the developer, in anticipation of the preparation of environmental impact documentation, or by those carrying out the assessment, as the impacts become clear, or by those responsible for granting the appropriate permits for the development, on receipt of the draft environmental impact statement.

While many environmental impact statements have led to the imposition of more stringent controls, EIA has also been utilised, at least in the United States, as a form of de facto land use planning leading to the amendment of the design of a development to make it more compatible with surrounding land uses. EIA, then, can be seen as a valuable weapon in the armoury of anticipatory controls over new stationary sources of air pollution, as its use encourages the employment of preventive technical and land use planning controls[3].

TECHNICAL CONTROLS OVER AIR POLLUTION

As shown in Figure 3, controls over air pollution may be imposed at various stages in the pollution process. For example, the production process may be altered to a less polluting one, ie, one in which fewer waste products are generated or in which the wastes are more effectively controlled. Such changes may be implemented either for purely process efficiency reasons or, more usually, because the imposition of environmental controls forces a reconsideration of the process to be made. Another method of control involves the amendment of the way in which goods are consumed (for example, by limiting the use of products) to create fewer wastes.

It is at the waste treatment stage that most technical controls are applied. The use of pollution control technology includes filtration, electrostatic precipitation, scrubbing, absorption and combustion techniques (Stern, 1977c). Briefly, filtration involves passing air through fabric or other types of filter to remove most of the entrapped particles. In electrostatic precipitators an electric field is used to charge particulate pollutants, attract them to a plate and cause them to fall out of the air stream. Scrubbing is usually used to remove gaseous particulates such as chlorides or sulphur oxides which are soluble in water. The plume is cooled substantially in this process and dispersion may be adversely affected. Absorption involves the removal of gaseous pollutants by trapping them onto sensitive surfaces. Activated carbon, for example, will absorb many malodorous gases. Finally, combustion involves igniting the gaseous pollutants in air at a high temperature to convert them from active chemicals to routine combustion products such as carbon dioxide and water vapour.

The method of control may be specified by the air pollution control agency, by insisting on certain types of equipment which meet specified performance standards, or it may be left to the polluter to choose that equipment which will meet the control authority's emission or air quality standard requirements. The control authority may also insist on various 'housekeeping' methods to control fugitive emissions, which are frequently a source of serious pollution problems. Dust suppression, by damping wind-blown material, covering and enclosing processes, equipment maintenance procedures etc, is frequently specified.

Several techniques of source control remain at the waste disposal stage. Local abatement can be achieved by maintaining the plume at high temperature to ensure that it is a given maximum lift, by discharging only at times when the atmosphere is not heavily polluted (meteorological control) and by making sure the emission takes place at a height well above that of nearby buildings to prevent the plume being caught up in eddies. Meteorological control is not widely employed but has been used in both the United States and in parts of Europe. Chimney height controls have been particularly popular in Britain and can have a profound effect on ground level concentrations (Chapter 5).

LAND USE CONTROLS OVER AIR POLLUTION

Land use planning controls apply almost exclusively to new or modified development, rather than to existing pollution emitters, and can be a valuable adjunct to technical and other controls at source. A given level of emission will cause quite different pollution problems if it occurs well away from sensitive receptors than if it takes place close to them in an area with poor dispersion characteristics. However, land use planning controls can never be as powerful as technical and other source controls or provide a substitute for them.

The role of land use controls

The land use planning agencies or authorities exert control at most stages in the pollution process (Figure 3) but their most powerful potential contribution is in determining the nature and location of new development and of redevelopment. Because pollution originates as waste from production and consumption activities, one of the key variables in pollution control – the geographical point at which additional waste is created – is determined once the location of these activities has been established. Therefore, because of their control over land use, planning agencies exercise an important influence on the spatial origin of wastes and consequently upon pollution levels and their distribution. These agencies are undoubtedly the principal controlling authorities in deciding the location of the pollution process, whether they recognise their position or not.

Control over the location of the pollution source is much more fundamental than other types of planning control over the pollution process. The new locations at which power is generated and at which goods are produced, and hence the locations at which the associated wastes arise, are largely determined by grant or refusal of land use planning permits.

The locations at which products are used can be directly controlled by planning authorities. Apart from allocating land for the consumption of goods (eg, residential areas), agencies have at least a voice in the determination of new road alignments (and also possess some indirect control over the use of existing roads), thus influencing air pollution arising from vehicular traffic. Further, and more fundamentally, by granting permission for certain types of seemingly relatively non-polluting development (such as sports stadia, commercial buildings and shopping centres) agencies are permitting so-called *indirect* pollution sources to arise as the large numbers of motor vehicles used in travelling to and from them will emit significant quantities of air pollutants. Land use planning authorities can also exert some direct control over the treatment of various wastes emitted from stationary sources by, for example, insisting upon particular air pollution emission levels (ie, requiring technical controls) or by specifying discharge height or by demanding certain building types for containment of pollutants.

The place at which the waste matter is disposed of is generally determined once a development is approved, although the precise location (and height) of, for example, a new chimney stack associated with the development may be subject to planning approval. Planning authorities have some control over waste diffusion, apart from the specification of stack heights or locations. They may, for example, insist on buffer zones and/or planting to remove pollutants from the atmosphere.

Land use planning agencies have a crucial role in controlling the damage arising from the resulting pollution, since they control the nature and location of receptors. In other words, apart from protecting the environment around a proposed new source of pollution, authorities can control damage from an existing source of pollution by determining the nature of new developments close to it. This may be achieved either through the granting or withholding of land use permits (eg, refusal of housing close to an oil refinery) or by the attachment of conditions (eg, that a school building be constructed so as to be separated from a major air pollution source by its playing fields).

It must be stressed that there are two stages in the planning process, the preparation of a plan and its implementation in the form of decisions on the use of specific areas of land. While all the controls mentioned above can be exercised in the absence of an overall land use plan, the potential role of the land use planner in ameliorating air pollution is not restricted either to attempting to ensure that the best anticipatory controls are imposed when development is permitted or to preventing such development. Rather, it extends to planning the future use of land to reduce air pollution by the preparation of implementable plans.

One final role must be mentioned. Apart from their controls at different stages in the pollution process, land use planning agencies are in a unique position – as a focus for consultation on both plan making and land use decision-taking – to play a central co-ordinating role in the control of pollution (Roberts et al, 1975; Wood, 1976).

In general, land use planning is not a particularly sensitive method of controlling pollution because planning decisions relating to a source tend to be inflexible. Unlike the systems of control which they complement, land use planners tend not to have such a continuing interest in, and control over, the pollution arising from a particular activity. Once a planning decision about a new source has been made, it can be altered only with considerable difficulty, no matter what technological changes affect the pollution arising from the source over the years. This lack of sensitivity does but little to detract from the usefulness of planning techniques in controlling air pollution is not disputed.

Land use control techniques

There are numerous techniques available to the land use planning agency for controlling air pollution from new stationary sources or for ameliorating the effects of existing air pollution on new receptors. These can be classified as:

1 siting of industry with respect to topography;
2 siting of industry with respect to sensitive receptors;
3 control of land use around sources;
4 use of buffer zones;
5 design and arrangement of buildings;
6 use of district heating.

The *siting of industry with respect to topography* is an important technique for controlling air pollution. Exposed, windy sites will allow maximum dispersion of air pollutants to occur and most commentators suggest that such locations, especially for low-level sources, should be chosen in preference to valleys and basins where pollution is liable to be trapped by temperature inversions (Craxford and Weatherley, 1966; Arnold and Edgerley, 1967). Chandler (1976) has summarised the position:

> Exposed hilltop and upland sites are preferable ... although detailed local analyses will have to be made of the frequency, location, depth and intensity of temperature inversions in relation to the point of emission, to make sure the plumes are not brought down by eddies set up by the topography or trapped beneath an inversion (p 46).

It is generally accepted that valley sites should be avoided wherever possible and that, if development of such sites is essential, very high chimneys discharging pollutants above the level of inversions are advisable (Scorer, 1972). If pollution sources are to be constructed in valleys, it is thought preferable to locate them on the windward slopes as wind-generated dispersion is greater (Rydell and Schwarz, 1968).

The *siting of industry with respect to sensitive receptors* such as residential areas, schools, hospitals, children's and old people's homes, and intensively used recreational facilities, is another effective technique. Separation is especially valuable in reducing the effect of particulate pollutants since these tend to fall out in a localised manner. Similarly, high stacks for the dispersal of gaseous pollutants may be employed with less aesthetic difficulty well away from receptors than adjoining them. There are obvious advantages in siting an offending industry at the centre of a large tract of ground to minimise concentrations at the periphery but the use of less polluting types of industry to surround the principal source is also effective (Voorhees et al, 1971).

The location of industrial pollution sources to the leeward side of a town is only really satisfactory when considering tall chimneys emitting large quantities of pollutants at high velocity. Despite the fairly common advice that such industries should be located so that the prevailing winds carry pollutants away from high concentrations of people (eg, World Bank, 1978: 48) this is not always an effective policy. For lower-level emissions wind speed is of more significance than wind direction and the worst pollution levels often accompany calm conditions or very light winds from non-prevailing wind directions (Chandler, 1976). The location of industry to the windward side of a town or city may, indeed, be preferable if the prevailing winds are strong and provide adequate dispersion of pollutants (Bach, 1972). The selection of industrial sites to minimise pollution concentrations should thus be a matter for detailed analysis (Branch and Leong, 1972).

If existing stationary sources give rise to pollution which cannot be effectively controlled it is possible to limit the effects of that pollution by *control of land uses around the sources*. Several measures are possible: ensuring that only industrial uses are permitted in the environs of the source; ensuring that only low density development of limited height is permitted in the affected area; providing planted buffers around the source in the form of open land; and reducing the population of the affected area (or preventing it from increasing) but using various spatial remodelling or density control measures (including demolition). Particular attention can be paid at this stage to the necessity to avoid locating very sensitive receptors (schools, hospitals, etc) close to the pollution source, although no specific distance standards are available (Voorhees et al, 1971).

The *use of buffer zones* between industrial uses and sensitive receptors follows from the notion of separation. Such zones are usually, though not always, kept free of development and may be dedicated to recreational use. Apart from producing a distancing effect (they emit no pollutants), open spaces planted with trees, shrubs and grasses alter local climate, increasing wind speeds and reducing temperatures, thereby encouraging air circulation and thus increasing dispersion of pollutants. Vegetation also directly absorbs pollutants on its foliage, thus reducing air pollution levels directly (Hill, 1971). While some exaggerated claims for the efficacy of green areas planted with trees in reducing pollution have been made (by, for example, De Santo et al, 1976) there is no doubt that the average concentration of a pollutant (particularly particulates) declines with increasing proportions of planted open space (Chandler, 1976). For example, concentrations of particulates at the centres of parks are normally much lower than at the margins. Beneath tree canopies the air contains a relatively low proportion of the pollution found above the foliage and in surrounding built-up areas.

There is conflicting evidence on the width of such 'sanitary clearance zones' necessary to purify the air but it has been suggested that 'buffer zones between basic industries and dwelling areas should, in many instances, have a width of more than 2km' (Craxford, 1976). This distance seems unrealistic in most urban situations and probably assumes only rudimentary technical control. Other distances of the order of 500m have been quoted (Hagevik et al, 1974:127) but smaller, planted zones can be surprisingly effective and are quite widely advocated (eg, Bach, 1972). Tree barriers between industrial and residential areas can thus reduce air pollution considerably, a plantation 30m deep giving a high degree of dust interception and significant reductions in gaseous pollutant concentrations.

Even a single row of trees can reduce particulate concentrations perceptibly if planted on a green verge (Saunders and Wood, 1974). Similarly, very small areas of open space in an urban area can reduce particulate pollution levels. Coniferous trees are more effective than deciduous trees in filtering out particulate pollutants as well as being evergreen and hence effective all year round. They are, however, more easily damaged, especially by gaseous pollutants (Bach, 1972). While the quantifiable advantages are not yet documented with any precision, the use of planted open spaces appears to be justified on air pollution control grounds, even if their size is very limited. (This is as true of planting in car parks as around buildings.)

The *design and arrangement of buildings* may have a considerable effect upon local concentrations of pollution since local temperatures and winds, the two principal determinants of atmospheric diffusion, are affected. Two aspects are involved: the effect of a new building or structure upon airflows and hence the distribution of pollution from other sources; and the effect of airflow around a building upon the pollution released by it (McCormick, 1971). While the intermixing of high and low buildings may elevate some sources and improve dispersion, there is a danger that emissions from low rise sources will cause high pollutant concentrations to affect the upper floors of high rise buildings.

To lower pollution concentrations, buildings should generally be sited to encourage the movement of air and to avoid street canyons (high buildings and relatively narrow streets), the creation of enclosed courts and 'V'-shaped arrangements which trap air pollutants. Lifting buildings on stilts and incorporating breaks in long building frontages also encourage circulation. Where possible verandahs, sitting areas, balconies and play areas should be sited away from pollution sources, particularly heavily trafficked streets. Similarly, structures in much frequented areas of the site should be set back from major roadways, but setbacks should be varied to increase turbulence and dispersion and thus reduce concentrations. The variation of building size and heights and the diversion of people from low areas on sites which trap pollutants will also ameliorate contaminant levels.

Pollution from residential areas, and to some extent from commercial areas, may be reduced by the *use of district heating*. District heating utilises centralised fuel-burning in a limited number of large, efficient and relatively easily controlled units, and can replace multiple low-level poorly-controlled sources of pollution. The arguments for district heating on pollution grounds are less convincing if relatively clean fuels such as natural gas are burned in dwellings.

Overall it is clear that the greatest impact on air pollution can be made by land use planning techniques when all the various methods are employed together, for which a land use plan will obviously be a necessary framework. However, there is very little reliable published quantification of the reductions in concentrations of different pollutants achieved, or even anticipated, as a result of employing the various land use planning techniques in controlling air pollution. The importance of land use controls in ameliorating pollution problems by themselves should not be exaggerated but they can be a valuable adjunct to technical and other source controls, as Kaiser et al (1974) have stated:

> Source control alone fails to take advantage of the assimilative capacity of the atmosphere which offers a possibility for spreading out emissions over a given area, thereby diluting their impact. Strategically locating sources may be a least-cost alternative to source control and may not conflict with economic objectives as greatly as source control regulations (p 382).

NOTES

1 Bureau of the Census (1986:193). These are global figures and include government and personal, as well as industrial, expenditure.
2 Brail (1975) and Venezia (1976) also mention district emission quotas and floating zone emission quotas, but these are essentially variants of emission allocation and emission density zoning.
3 This is certainly the view of the Commission of the European Communities (1977) which, in its second action programme, made EIA the central plank of its measures to encourage the anticipation and prevention of environmental problems.

3
Preventive Controls over Air Pollution in the United States

This chapter describes the context of, the legal and administrative arrangements for, and practice in, the anticipatory control of new stationary sources of air pollution in the United States. It is divided into four parts. The first describes the philosophy of environmental regulation and some of the factors which determine it. The next part of the chapter is concerned with air pollution. The main air pollution trends are briefly presented. There follows an outline of the legal framework for air pollution control and, in particular, for stationary source control. Practice in utilising these provisions is then discussed. An exposition of land use planning control follows. Third, the legal framework of local and state land use control is outlined, and practice in utilising these controls is explained. Finally, the role of environmental impact assessment in controlling air pollution is reviewed.

CONTEXT

Physical characteristics

Perhaps the most obvious difference between the US and the UK is the relative size of the two countries. The United States has an area of nearly 10,000,000 square kilometres on which to house its population of over 240,000,000 (Bureau of the Census, 1986:8). While this population is growing at over 1% per annum, the average population density, at 25 persons per square kilometre, is very low. In 1985, over 75% of the population of conterminous USA (the 48 states, excluding Alaska and Hawaii) lived in metropolitan statistical areas (MSAs) or consolidated MSAs, occupying 16% of the total land area (pp 26–27).

Geographical differences account for the divergence of the United States from many of the social traditions of its erstwhile colonial master. As Williams (1960) put it, the *distance* between the US and Europe:

> minimized the danger of large-scale invasion by major foreign powers and contributed to a certain fluidity and 'openness' in social relations and to the

development of a decentralized, nonsecretive, nonmilitary, governmental structure.

The existence of a *frontier* meant 'expansion, opportunity, economic growth, social mobility' and the great expanse of a *contiguous land mass* gave 'a loose and amorphous quality to the total society, while abundant natural resources guaranteed a certain degree of independence (pp 8–9).

The seemingly limitless land area led to the American myth of superabundance of land and natural resources. The legacy of the early settlers has been the pioneer mentality in which nature was seen as something to be fought, not something to work in harmony with. This *frontier ethic*, in which every individual seeks to live in his own house on his own (preferably large) piece of land, and to use that land as he chooses, is an abiding part of the American psyche. (It has now been joined by a desire to own a motor car and use that as he chooses.) Land is generally seen, therefore, as a commodity to be possessed, exploited and conquered. While there are indications that this ethic is slowly beginning to change[1], it is important in explaining the fundamental reluctance of politicians to limit personal freedoms in the use of land. It also helps in comprehending the capitalist values of American society and why America has remained very much a business culture.

Somewhat surprisingly, perhaps, as the Council on Environmental Quality (1982a:215) reported, the federal government owns about 33% of the land in the United States, much of it in unproductive areas of the western states which were never claimed by settlers. The states and local jurisdictions own another 6% of the land. Land ownership (rather than land use controls) has been, to a large extent, the preferred method of ensuring the 'appropriate use' of land in the US. The national parks are the outstanding example of this power to exclude unwanted land uses from areas managed in the public interest. Because much of the most scenically attractive land in the US is in public ownership, the pressure for land use controls on the remainder has been less than might otherwise have been the case. Another reason for the absence of widespread demand for effective controls is probably that much land is neither farmed nor in urban use: it is simply not utilised, even in apparently urbanised regions like the north east (Clawson and Hall, 1973). Generally the development of such land is unlikely to provoke conflict.

The physical characteristics of the United States may help to explain the concern with means rather than ends which is so characteristic of the country. Kouwenhoven (1961) advanced the thesis that 'America is process':

> Our history is the process of motion into and out of cities; of westering and the counter-process of return; of motion up and down the social ladder – a long, complex, and sometimes terrifyingly rapid sequence of consecutive change. And it is this sequence, and the attitudes and habits and forms which it has bred, to which the term 'America' really refers (p 72).

This concern with process or means can be seen in the lack of emphasis in planning on fixed target populations or on the physical distinction between city and country, and in the organisational arrangements for land use control, in zoning ordinances and in the complex air pollution regulations in the USA.

The United States is cross-cut by religious, ethnic, racial and regional differences. For example, about 12% of the total US population is black, of which the majority is of Hispanic origin (Bureau of the Census, 1986:35). The various waves of immigration and regional differences mean that there are numerous interest groups which any national legislation must take into account, each having its own goals. The US is a classical pluralist or 'individualistic' (utilitarian) society (Meyerson and Banfield, 1955), in which there is little pretence at seeking common interest and consequently, considerable suspicion of all big organisations, including big business. The design of environmental protection regulations must, perforce, leave little discretion to the individual.

Climate

The climate of the United States exhibits marked variations, both geographically (from the deserts of Nevada to the lakes of Maine, from the southern sunbelt to the northern frostbelt) and seasonally (temperature ranges of well over 50°C are common).

A common feature of the United States' climate is the propensity in many cities towards high summer temperatures and stagnant air. This, together with the bordering of many western cities by mountains that interfere with air movements, leads to the frequent formation of inversions. An upper, cooler layer of air prevents the dispersion of the rising air with its burden of pollution, which builds up until the inversion breaks down. High levels of car ownership and usage thus lead to serious photochemical smogs of the type characterised by, but by no means confined to, Los Angeles. Further, air pollutants exported from one region tend to cause damage in other regions of the USA, as well as in Canada. Acid rain is regarded as a real threat to the environment and air pollution is seen as a serious political issue.

In the United States, as in Britain, widespread changes in heating patterns have taken place (including changes to wood burning with consequent smoke pollution in Oregon, Colorado, and other states). However, the general perception is that pollution is industry's affair, not the people's. Thus the solution to automobile exhaust pollution is seen not to be vehicle restraint but stricter initial controls, preferably without subsequent inspection and maintenance programmes. This view that pollution is caused by industry is coupled with a feeling that many pollutants are unsafe at any concentration and naturally colours opinions about the siting of new sources of air pollution.

Government system

The United States has essentially three levels of government: federal, state and local. The state is the fundamental source of sovereignty in the United States. The federal government is responsible for such matters as defence and foreign policy and for overseeing the provision of services and the imposing of controls by the states. This oversight is achieved mainly by passing legislation, subsequently formulating regulations and providing funding. The states each have their own legislatures, generally consisting, like the federal government, of upper and lower houses and a governor (a state equivalent of the head of the executive branch of the federal government, the president). Each state also has its own judiciary.

Generally, pollution control responsibilities are shared between federal and state levels, with the states usually being responsible for implementation. Thus, each state is responsible for air pollution control, in conjunction with the regional offices of the federal Environmental Protection Agency (EPA). EPA was formed in 1970 to centralise the control of environmental problems.

There are numerous local governments and special purpose authorities. Indeed, it is remarkable just how numerous these counties and municipalities are, and how little has been done to reform them. In 1982, the United States contained 3041 counties (of which 725 had populations of less than 10,000), 19,076 municipalities (of which 16,879 had fewer than 5,000 people), 16,734 townships (of which 15,715 had fewer than 5,000 people), 14,851 school districts and 28,588 other special purpose districts, making a total of over 82,290 local governments, an increase of 2,428 since 1977 (Bureau of the Census, 1986:249). Each of these governments has its own officials and elected representatives. They vary in population from a handful to seven million and in size from a few city blocks to 20,000 square miles.

Although many states undertake some form of land use control, the major regulation of land use is normally carried out by county, municipal and township governments. These local authorities are far more independent of state and federal governments than their British equivalents and they gain directly from attracting new industry because local property taxes accrue to them alone. It is, in the circumstances, unsurprising to note that local government corruption is by no means uncommon.

There has long been recognition of the rapid growth of the metropolitan areas and the increasing spread of population across existing jurisdiction boundaries, with a polarisation between the working classes (frequently coloured) remaining in the central city and the middle classes living in suburbia. There have similarly been many attempts to reform local government, most of which have failed because the political advantages of the status quo are too attractive to the constituencies. There have been some notable exceptions: New York City in the early years of the twentieth century and the Greater Miami Metropolitan Area in 1957 – though this latter is still a relatively weak coalition (Greer, 1962). There appear to be three main reasons why the bewildering complexity of metropolitan government remains:

> These are (1) the underlying cultural norms of Americans concerning local government, (2) the resulting legal-constitutional structures, and (3) the political-governmental system built upon them ... Based upon a deep distrust of governmental officials and a faith in the competence of the ordinary voter, they also lead to such corollaries as 'that government is best which governs least' (and more important *costs* least) (Greer, 1962:124, emphasis in original).

Recently, the Reagan administration has handed more responsibilities back to the local governments, while not increasing their budgets, causing problems of cuts or revenue raising in many areas.

Because of the difficulties of re-organising local government, a quasi-regional level of government has sprung up. This is known as the council of governments or the regional planning commission. Its principal purpose was to try to overcome the inefficiencies of local governments, which:

> Suffer from multiple, overlapping jurisdictions, bureaucratic proliferation, and wasteful duplication of local-regional-state-federal-agencies, each presumably designed to resolve the same problem but competing for increasingly hard-to-get public funds (Mogulof, 1971:vi).

There are now councils of governments for almost every one of the MSA's. However, recent actions by the Reagan administration have cut the budgets of the councils and both their functions and their staff have been severely pruned.

In the United States, big government has been viewed with just as much suspicion as big business (and big unions) and the modification of urban electoral procedures through nonpartisan ballots, referenda and recall procedures have been seen as the way to revive individualism, by giving every man an equal chance to govern (Heidenheimer et al, 1975:112–113). This suspicion has led to the creation of the freedom of information provisions in the United States, together with formal rights of objection and appeal against many government decisions. This extends considerably beyond the right to be elected or nominated to government bodies and includes rights of speaking at hearings into, and third party appeals against, air pollution control and land use planning decisions (Callies, 1981). While some participation may be little more than a meaningless formality, it can provide a valuable check on official abuses. The government system in the United States, with its naked self-interest and parochialism, in which the broader view is a preparedness to compromise by removing a block on action at the last possible moment, can almost be characterised as organised anarchy.

The United States not only has numerous government agencies, it also has, because of the desire to codify the separate state legal systems, numerous laws and, as a consequence, a multitude of lawyers. It is reputed to have more lawyers per 1000 inhabitants than any other country in the world and is remarkably litigious.

The traditional concern of each individual to participate in his own government, together with the sheer size of the United States, and the capitalist culture, probably help to explain the large number of local papers, local radio stations and local television stations across the country. These are frequently much more confined in their geographic coverage and more numerous than in the United Kingdom. They consequently tend to give far greater attention to, and to be more likely to take positions on, local environmental controversies than their British equivalents.

There is one further relevant consideration in explaining the high level of concern about environmental affairs in the United States. The income tax system encourages many higher income earners to donate money to charitable organisations like the Sierra Club. This, together with a career structure which encourages participation by the able (at least for a period) in environmental issues, ensures that pressure groups can often mount thorough, well-researched and well-argued opposition to a wide range of new pollution sources.

Environmental regulation

As a result of its national characteristics the United States has developed a centralised regulatory system of environmental standards based on specific criteria (eg, health impacts) with explicit administrative procedures. This heavy reliance on formal rules and procedures has the advantage of being visible and being subject to public participation but frequently entails excessive legal confrontation and is cumbersome.

There has been an explosion of environmental regulation in the United States since the 1960s (Bosselman et al, 1977). New regulations and regulatory agencies have been added to old ones at federal, state and local level. Noble et al (1977) stated that there were some 137 federal programmes having a direct impact on land use, including the Clean Air Act. States have acquired regulatory powers over numerous types of land use (eg, power plants) and have set their own environmental standards (eg, air quality levels). Thousands of local governments have established control programmes to achieve goals such as environmental protection and growth management.

This process has been mainly one of addition rather than substitution and developers have had to adjust to the task of reconciling the satisfaction of not only multiple objectives, but of conflicting institutions (since these agencies are each independently capable of vetoing development proposals). Not only have regulations changed rapidly in recent years but the staff of the various agencies administering them has also been subject to very high turnover. As a result of public involvement in many of the processes, there have been many thousands of environmental law suits. The transaction costs of winning a permit to develop have obviously escalated.

Environmental policy has been criticised from two directions: environmentalists accuse it of being ineffective, while industrialists, and others concerned with the costs of regulation, attack it for being inefficient (Vogel, 1986). Indeed, a fierce and controversial debate about environmental regulation has been waged in the United States.

There is clearly a regulatory problem in the United States. A general consensus exists that the environmental protection process is cumbersome and time-consuming and that some regulations, especially local rules, are not cost-effective (eg, Fix and Muller, 1982). As one EPA (1982) report put it, citing the 715 permit requirements for the SOHIO pipeline from southern California to Texas which was ultimately abandoned:

> The complex of environmental laws that exists today was formed incrementally over time: each new law was passed to address a specific single purpose or need, and subsequent laws were passed to fill in gaps left uncovered by the old. Moreover, organisationally separate agencies and programs also developed incrementally at the local, state and federal levels. As a result of this history, these agencies frequently have overlapping, duplicative, or contradictory regulatory authority as well as inadequate communication networks (p 3).

Bardach and Kagan (1982), for example, have argued that policymakers, prodded by politicians and public interest groups, have indulged in over-regulation that has generated unreasonable and inflexible regulatory behaviour which, in turn, has bred resentment and resistance in industry.

There has been an increasing willingness on the part of EPA to utilise measures which give business the opportunity to decide how to achieve reductions in pollution itself, so long as amelioration is achieved. It is encouraging industry to follow a *command, counterproposal and control* regulatory path, rather than the traditional *command and control* route (eg, Drayton, 1981). It is thus going some way towards meeting the suggestions of economists for greater flexibility and for the use of taxes on residuals in controlling pollution (Friedlaender, 1978). It has also encouraged the streamlining of the state regulatory process (EPA, 1982).

There is a distinct lack of trustworthy evidence on the actual locational effects of environmental regulation on industry. However, Duerkson (1983), found little to support many of the criticisms:

> The right to pollute is not an important locational determinant. No evidence of

migration of industry from one state to another in search of 'pollution havens' was unearthed . . .

... a significant number of industrial facilities have been built relatively quickly over the past decade with few or no serious environmental problems. These success stories are often overlooked in the clamor over celebrated siting battles.

... a good deal of the regulatory delay of the 1970's can be attributed to 'teething pains' that are likely to be eased as the players in the siting game learn the new rules (pp xxx-xxii).

He found no evidence, in his exhaustive study of siting problems, that environmental regulations were causing a flight of industry abroad or that other countries were better at reconciling environmental regulations with industrial development.

Among other researchers endorsing this view, Storper et al (1981) came to much the same conclusion: because of local eagerness to attract new growth, 'industrial corporations ordinarily make their siting decisions on economic grounds and deal with regulation as a secondary consideration'. Similarly, Stafford (1985) reported that his study of company location did 'not support the contention that environmental regulations will lead to major shifts in the location of industry within the US.' Again, Leonard (1982) concluded that 'the costs of complying with environmental regulations are not emerging as a decisive factor in most industrial decisions concerning desirable plant locations . . .' It would thus appear that many of the accusations levelled against environmental regulations have been inaccurate or exaggerated.

Healy (1982) observed that environmental blockage of greenfield plants may have had the salutary effect of channelling investments toward modernisation or expansion at already industrialised sites. He further stated that for several 'pollution intensive sectors' new investment would probably result in net environmental gains and suggested the possibility of a less adversarial relationship in the future between environmentalists and industrial developers.

Although the Reagan administration has made strenuous attempts to reduce environmental regulation, these have proved markedly unsuccessful and, indeed, the power of environmental lobbies has increased rather than decreased (Symonds, 1982), resulting in significant senior personnel changes. However, there appeared to have been a marked decline in the standard of enforcement of regulations in the early 1980s, largely as a consequence of budget cuts which affected federal, state and local environmental control agencies (Conservation Foundation, 1982). The Reagan era has thus been characterised, overall, by marking time on environmental regulation and by an increased interest in compliance with existing regulations (Conservation Foundation, 1984:10). It has not, of course, been a period of industrial expansion and much manufacturing development has been achieved by extension and modification, rather than by the utilisation of greenfield sites.

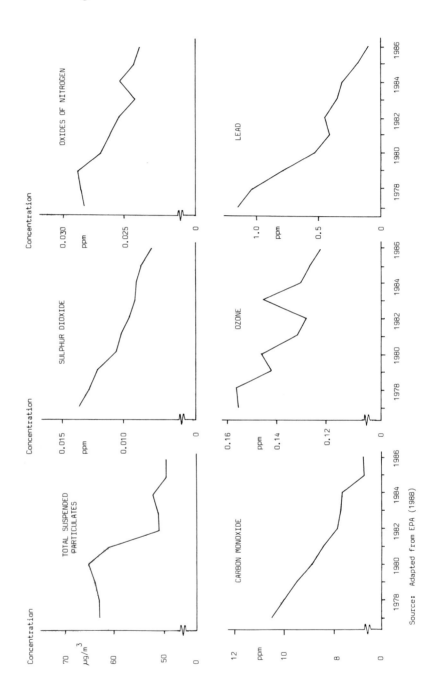

Figure 4 US air quality trends 1977–1986

AIR POLLUTION CONTROL

Trends

The United States has experienced a significant improvement in air quality since 1970. There have been further improvements since the Council on Environmental Quality (1981) reported that:

> Air quality has been improving. The number of air pollution alerts has been reduced. The large amounts of chemical pollutants coming from industry have been reduced, and emissions from the worst offenders have been markedly curtailed.
> But much remains to be done in cities and elsewhere. There is growing evidence that wilderness, parks, and other pristine areas may be threatened by air pollution (p 271).

Figure 4 shows the ambient pollution trends over the last few years. It can be seen that a 25% drop in total suspended particulates during 1977–1986 resulted in a 23% fall in concentrations in the same period (EPA, 1988). This was probably due to the prevalence of fugitive emissions (not emitted through stacks) and windblown dust, rather than stack emissions, in total arisings (Conservation Foundation, 1987). On the other hand, a 21% fall in emissions of sulphur dioxide during that period was more than matched by a 37% fall in the national ambient sulphur dioxide level.

Emissions of oxides of nitrogen fell by about 8% between 1977 and 1986, and concentrations fell by 14%. While emissions of carbon monoxide dropped by about 26%, concentrations fell by 32%. Ozone concentrations decreased by 21% over the same period and the number of days during which the ozone air quality standard was exceeded declined by 38%. Emissions of volatile organic compounds fell by 19%. Finally, lead concentrations fell by 87% between 1977 and 1986, reflecting a drop in emissions of 94% over the same period (EPA, 1988).

Much of this improvement can be attributed to the Clean Air Act 1970. As the National Commission on Air Quality (1981) has stated:

> It is impossible to determine precisely the extent of additional air pollution that would exist if Congress had not enacted the Act, but it is certain that the quality of the country's air would be far worse now than it was in 1970, rather than substantially better (p 2.1–17, 18).

Legal framework

The United States has suffered from many air pollution problems, including the notorious incident at Donora, Pennsylvania, in 1948 when a large number of people were taken ill and several excess deaths were recorded, but none appear to have provoked a direct legislative response of the type observed in Britain. Concern about air pollution has a lengthy history: several cities enacted their own air pollution control legislation in the 19th century and 23 of the 28 largest cities had passed such laws by 1912 (CEQ, 1970; Melosi, 1980). The first federal legislation was passed in 1955, and the Clean Air Act appeared on the statute book in 1963. This act only assigned the function of identifying harmful pollutants to the federal government, leaving action to the states, but the 1967 Air Quality Act authorised the federal government to develop standards and implementation plans for states that failed to adopt these. The results of this act were disappointing, mainly due to the fact that the major responsibility for implementation had been left to the states with little federal coercion. The 1970 Clean Air Act, passed in the heady days of peak environmental concern, was much stronger and laid the main burden for federal oversight on the newly created Environmental Protection Agency.

EPA was required to develop and promulgate national ambient air quality standards for various pollutants (based upon health effects) to protect the most sensitive segments of the population. The country was divided into about 250 air quality control regions and the states had to prepare *state implementation plans* to meet the air quality standards in these regions by a set date. EPA had the power to take over this task if the states failed to respond and to impose sanctions where necessary. The Act specifically provided for citizen enforcement through the courts. Crucially, the standard-setting process was not to be delayed or watered down by cost or other non-health considerations.

The Clean Air Act of 1977, while not departing significantly from the 1970 provisions, reflected the new concerns about the cost of regulation and the energy crisis. This act attempted to replace the health-only rule of 1970 with one which balanced economic factors with the need to protect public health and welfare, though without altering quality standards. It was about to be reauthorised at the time of writing, and further major changes can be anticipated. The 1977 Act provides a comprehensive statutory framework, dealing as it does with pollution from motor vehicles (requiring certain emission standards to be met on new cars and to be sustained through an inspection and maintenance programme), with transportation control plans, with hazardous pollutants, with visibility protection[2], and with the preparation of state implementation plans. It also, of course, contains provisions relating to emissions from existing sources of air pollution, as well as from new or modified stationary sources of air pollution[3].

EPA has set and subsequently somewhat modified national air quality standards for six pollutants, and designated several hazardous pollutants, in order to protect human health and welfare. Sulphur dioxide and particulate matter less than 10μm in diameter (previously total suspended particulates) originate mainly from stationary sources and carbon monoxide, ozone, nitrogen dioxide and lead mainly from motor vehicles. There is no longer a standard for hydrocarbons. The standards for these *criteria* pollutants are summarised in Table 2 (EPA 1988). The primary air quality standards are based on human health effects and the secondary standards on damage to public welfare, vegetation, property, scenic value, etc. The dates set for the attainment of both sets of standards have been progressively postponed.

Stationary source control

To say that the United States provisions relating to the control of air pollution from stationary sources are arcane would be too generous. They are so labyrinthine that it has been claimed that no one understands fully both the Clean Air Act and the various regulations promulgated to implement it[4]. The following summaries provide brief explanations of the new source performance standards, prevention of significant deterioration and offset requirements of the new source review programme: (Raffle, 1978; Quarles, 1979; Environmental Research and Technology, 1982; Callies, 1984; Liroff, 1986). Any given source may be subject to all three types of regulation.

Most of these requirements have to be written into the state implementation plans, which may also contain additional new stationary source control requirements. Each state has promulgated plans which contain procedures for reducing pollution from existing stationary sources by

TABLE 2 US National ambient air quality standards

Pollutant	Standards	
	Primary	Secondary
Particulate matter (PM_{10})	Annual (geometric mean) 50 $\mu g/m^3$ 24 hour: 150 $\mu g/m^3$	Annual (geometric mean) 50 $\mu g/m^3$ 24 hour: 150 $\mu g/m^3$
Sulphur dioxide (SO_2)	Annual (arithmetic mean) 80 $\mu g/m^3$ (0.03 ppm) 24 hour: 365 $\mu g/m^3$ (0.14 ppm)	3 hour: 1300 $\mu g/m^3$ (0.5 ppm)
Nitrogen dioxide (NO_2)	Annual (arithmetic mean) 100 $\mu g/m^3$ (0.05 ppm)	Annual (arithmetic mean) 100 $\mu g/m^3$ (0.05 ppm)
Carbon monoxide (CO)	8 hour: 10 mg/m^3 (9 ppm) 1 hour: 40 mg/m^3 (35 ppm)	– –
Ozone (O_3)	1 hour 235 $\mu g/m^3$ (0.12 ppm)	1 hour 235 $\mu g/m^3$ (0.12 ppm)
Lead	3 months: 1.5 $\mu g/m^3$	3 months: 1.5 $\mu g/m^3$

Source: EPA (1988)

requiring the use of *reasonably available control technology* (RACT) and/or *bubble* provisions. In this latter approach the whole of a plant, factory or complex may be considered to be a pollution source, rather than the individual sources of emissions within the whole, leaving the operator free to choose how best to implement controls. These need not be of a purely technical nature but might, for example, involve process changes (Chapter 2). State implementation plans, most of which have been repeatedly revised, may contain air quality standards which are more (but not less) stringent than those required by the federal government.

Any new or modified source will be subject to the requirements of the state implementation plan. This normally specifies both limitations on allowable emissions and procedural requirements for preconstruction review. Thus, nearly all new pollution sources require prior review before operation can commence. This is true even for changes of use of existing buildings or for changes of ownership of a continuing operation. Construction and operation permits are usually required, the latter being renewable periodically. The decisions made by the relevant air pollution control agency are generally open to public participation, sometimes in the form of a hearing. The anticipatory arrangements for control in the United States are thus comprehensive, virtually every new or modified source being subjected to a preconstruction review of some type.

New source performance standards
Quite apart from any individual requirements in the various state implementation plans, EPA is required to set national new source performance standards (NSPS) for individual industrial categories. These require new plants to utilise the best system of emission reduction that the agency determines has been adequately demonstrated. If an NSPS has been issued, or proposed, for a particular type of plant, EPA regulations impose requirements that the owner must give advance notification to the state before beginning construction, with further notification due before actual start-up[5]. There is no minimum source size limitation on the relevant requirements.

New source performance standards have now been promulgated for pollutants emitted by over 50 types of industrial facilities (including incinerators, cement plants, petroleum refineries, iron and steel mills, etc)[6]. These are effectively national standards for all new stationary sources in these classes, reflecting the degree of emission limitation and percentage reduction in emissions achievable taking into account control costs, the health and environmental consequences of emissions reductions and energy requirements (National Academy of Sciences, 1981; Liroff, 1986).

Prevention of significant deterioration
The prevention of significant deterioration (PSD) programme applies to areas of the country which are already clean enough to meet the ambient air quality

standards. The provisions were included in the 1977 Clean Air Act to prevent a possible flight by industry from the polluted areas to areas where little or no previous development had occurred, with a risk of downgrading the pristine air of such areas. There is an area classification scheme, in which most national parks, national monuments and national wilderness areas are designated as Class I. The rest of the country (including industrial regions) is designated Class II (areas of moderate growth) with complicated and onerous requirements should states wish to reclassify areas as Class III (areas of major industrialisation) or, for that matter, as Class I[7]. To date there have been only one or two reclassifications from Class II to Class I, but there are still no Class III areas.

The Clean Air Act 1977 established *increments*, the numerical definition of the amount of additional pollution which may be allowed through the combined effects of all new growth in a particular locality. These are shown in Table 3 and are specified both for short and long time periods for the two stationary source pollutants: sulphur dioxide and particulates[8]. The effect of the increments is to create a whole set of de facto air quality standards varying throughout the country, since the same increments added to varying background levels (fixed in or after 1977) yield varying limits, though none may exceed the national ambient air quality standards.

Because the available increments might be utilised by the first firms moving into an area, EPA specified that each major new plant must install the *best available control technology* (BACT) to limit its emissions of those pollutants exceeding certain annual tonnages (see later). The statute specifies that energy, environmental and economic impacts and other costs must be taken into account in specifying BACT, which must be determined on a case by case basis. (It is thus not dissimilar from the British *best practicable means* for an individual works.) The BACT requirement is at least as stringent as an applicable NSPS.

To implement these controls, EPA imposed a requirement that each new

TABLE 3   PSD increments for air quality classes

Pollutant		Maximum allowable increase ($\mu g/m^3$)		
		Class I	Class II	Class III
Particulate matter	Annual geometric mean	5	19	37
	24-hour maximum	10	37	75
Sulphur dioxide	Annual geometric mean	2	20	40
	24-hour maximum	5	91	182
	3-hour maximum	25	512	700

NOTE: For specified non-annual periods (eg, 24-hour, 3-hour) the allowable increment may be exceeded during only one such period per year at any receptor site.

source should undergo a preconstruction review. It prohibited a company from commencing construction until this review had been completed and demanded that, as part of the review procedures, public notice should be given and an opportunity provided for a public hearing on any disputed questions of fact.

The Clean Air Act requires twenty-eight industrial categories of plant to meet the PSD requirements if potential emissions (ie, the maximum capacity of a source to emit pollutants under its actual physical and operational design, which includes any air pollution control equipment) of any regulated pollutant (ie, the pollutants for which air quality standards apply or the hazardous pollutants) exceed 100 tons per year. The plants specified include large municipal incinerators, petroleum refineries and some of the other types of source included in the NSPS listing. In addition, a new plant in any other category is also covered if its potential emissions of any regulated pollutant would exceed 250 tons per year.

If a new source is to be subject to the PSD requirements, the preconstruction review includes:

1 a case-by-case determination of the controls required by BACT;
2 an ambient impact analysis to determine whether the source might violate applicable increments of air quality standards;
3 an assessment of effects on visibility, soils and vegetation;
4 submission of monitoring data;
5 analysis of air quality impacts projected as a result of growth associated with the new facility;
6 full public review.

It should be noted that the impact analysis is based upon a modelling procedure using assumed ambient baseline concentrations (because certain sources such as construction activities are not taken into account in the calculations but new sources about to come on-line are included) rather than actual measurements of current air quality. The models endorsed by EPA tend to be conservative, ie, they overestimate the effect of a new source on the increments. It is up to the air pollution control agency to keep track of changes in emissions (eg, from minor stationary sources and road traffic) and the consumption of the increments (National Academy of Sciences, 1981).

Modifications to major existing plants require a PSD permit if the net emissions increase due to the modifications will exceed the values on a minimum emissions list, called a *de minimis* list (100 tons per year of carbon monoxide, 40 tons per year of oxides of nitrogen and sulphur, and of volatile organic compounds, and 25 tons per year of particulate matter). A modification to an existing minor source would not require a PSD permit unless the proposed addition would be major in itself. If the predicted expected ambient impact at locations outside the plant property of increases in emissions from modified sources are lower than the values on another *de minimis* list, this time of significant air quality deterioration limits (for example, 24 hour levels for particulates and sulphur dioxide are 10 and 13 µg/m³ respectively), an exemption may be granted from the requirement to obtain the preconstruction monitoring data which normally has to be provided.

The elements of this procedure are set out in Figure 5 (Environmental Research and Technology, 1980). It perhaps needs to be added that there are intense disputes not only over the types of models to be employed, but over the air quality standards themselves. The determination of BACT, too, is fraught with difficulties relating to the definition of just what is reasonably practicable.

The PSD programme is intended to be carried out by the state through the mandated state implementation plans and most states now have approved plans and delegated authority to administer all the federal regulations. PSD regulation has frequently been undertaken, in whole or in part, by the ten regional offices of EPA. This duality of control has merely confused what is already a complex and onerous permitting process, as Figure 5 amply demonstrates.

Nonattainment
If the PSD provisions set the requirements for new projects in clean air areas, the nonattainment provisions of the Clean Air Act apply to the dirty air areas which have failed to attain compliance with the ambient air quality standards. Where they apply, and they apply in most of the industrialised parts of the country, the nonattainment provisions are even more restrictive than the requirements of the PSD programme[9]. They are that:

1 the new source must be equipped with pollution controls to ensure the *lowest achievable emission rate* (LAER), which in no case can be less stringent than any applicable new source performance standard:
2 all existing sources owned by an applicant in the same region must be in compliance with applicable implementation plan requirements or under an approved schedule or an enforcement order to achieve such compliance;
3 the applicant must come up with sufficient *offsets* – reductions in emissions from other existing sources – to more than make up for the emissions to be generated by the new source (after application of LAER);
4 the emission offsets must provide a 'positive net air quality benefit in the affected area' (Quarles, 1979:16).

The LAER requirement demands individual determinations as to what emission control technology can achieve in each case. This is a complex and uncertain process, usually relying on the determination of the most stringent technology in commercial operation for at least a year. LAER is a technology-forcing standard, since energy use and costs are not supposed to be

48 *Planning Pollution Prevention*

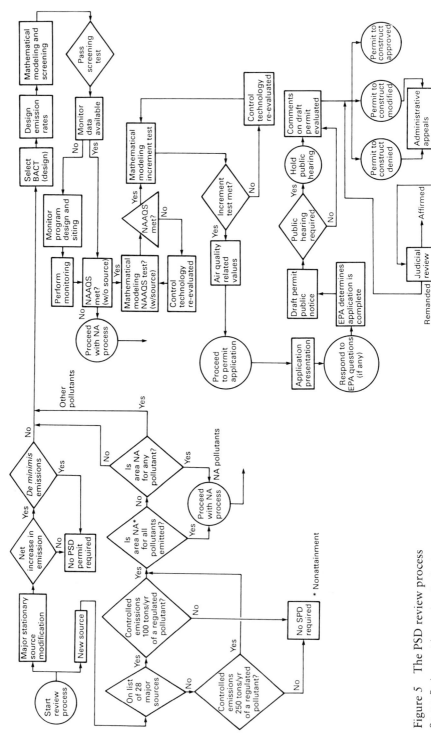

Figure 5 The PSD review process

Source: Environmental research and technology (1980)

considered in specifying it. As well as never being less stringent than the relevant new source performance standard, LAER can never be less stringent than the best available control technology requirement.

The offset provision is novel. As Liroff (1980) has explained:

> The trading of pollution offsets allows industries to site or expand in areas currently violating national ambient air quality standards established by the Clean Air Act (p 1).

Offsets must represent emission reductions, from the developer's own plants or from other plants, which would otherwise not be required and they must be greater than one-for-one. There is a great deal of discretion in the way in which an agency carries out its offset accountancy, but the offsets must be legally enforceable. The net air benefit requirement has generally been taken to mean that there would be no serious detriment anywhere due to the new source. This frequently involves modelling the anticipated concentrations, utilising data on existing concentrations and on other emissions in the area, in a similar way to PSD simulations.

Nonattainment provisions apply to new and modified facilities which have the potential to emit 100 tons per year of particulates, sulphur dioxide, nitrogen oxides, carbon monoxide or volatile organic compounds. LAER, offsets and related conditions do not apply to smaller plants. However, if certain equipment in an existing plant is replaced, the new equipment may be subject to the nonattainment requirements (including the obligation to install LAER) even though there is a net reduction in total emissions from the overall plant. The same emission limits apply to modifications as to new plants. There are provisions for banking offsets on both a formal or informal basis, and a limited number of banks has been set up (Liroff, 1980). Bubbles, offsets and banking have now been incorporated into one emissions trading policy, which applies to new sources as well as existing sources (Liroff, 1986).

Practice and reform

The public in the United States has persistently supported controls on environmental pollution generally, and on air pollution in particular, despite general support for reducing regulatory burdens on industry. A 1985 survey, confirming earlier results, found that 73 per cent of the respondents interviewed favoured enforcing pollution regulations 'even if workers might be laid off as a result' (Conservation Foundation, 1987:21). The Clean Air Act has been at the centre of continuing debate over the degree of environmental regulation to which business should be subject. In particular, the national ambient air quality standards, the new source performance standards, the nonattainment area offset requirements and the prevention of significant deterioration requirements have all attracted criticism.

Although EPA has numerous supervisory or oversight responsibilities, the Clean Air Act is basically administered by the states, some of which delegate some powers to certain regions or to counties and municipalities. The states, rather than the more local agencies, are usually responsible for stationary source controls. However, many states have felt that there is too much duplication of effort between the EPA regional offices and the state air pollution control agencies and even that federal pre-emption of state prerogatives has sometimes occurred. It is, nevertheless, widely recognised that controls over new sources are effective, though controls over existing sources frequently leave much to be desired.

The necessary air pollution control legislation to implement the Clean Air Act has now been passed in every state but progress on state implementation plans varies. The precise administrative arrangements for air pollution control differ. Some states use health departments, some have created *little EPAs*, and many have so-called *super agencies* (very large departments having a wide range of responsibilities). Citizen involvement also varies between states. In general, the air pollution agencies tend to be somewhat remote from public control, though not so remote as Her Majesty's Inspectorate of Pollution in Britain.

Various estimates of the cost of implementing the Clean Air Act, including the costs of tax concessions on new control equipment, have been made. Apart from the Bureau of the Census (1986:193) figure of about $30,000M for 1984, the National Commission on Air Quality (1981) has also made estimates of the costs of control. The total cost of capital expenditure in 1978 was put within the range of $1800–7500M with operation and maintenance costs being rather more (p 4.1–4). Figures published by the Council on Environmental Quality (1984) indicate that total spending on air pollution abatement and control in 1981 was $29,494M.

No estimates of the total number of people involved in air pollution control appear to exist. However, it is known to be very large. The manpower involved was of the order of 16,000 in the Environmental Protection Agency, in state agencies and in local government agencies in the late 1980s (Air Pollution Control Association, 1988; Conservation Foundation, 1984). This, of course, excludes those employed in air pollution control within the private sector. Fines levied by air pollution control agencies can be very substantial. It is not uncommon to find states levying a total income from fines of over a million dollars per annum, with individual fines sometimes extending to hundreds of thousands of dollars.

The National Commission on Air Quality (1981) found that many state and local air pollution agencies appeared ill-equipped to accept substantial additional responsibilities without further financial resources and personnel. It reported that these agencies were not demanding LAER because of inadequate resources and information exchange, that offsets (at least *paper offsets*) have generally been available and that companies were reluctant to *bank* emissions reductions because they wished to retain them for their own possible future expansion.

In relation to industrial expansion, the Commission stated:

> The nonattainment and prevention of significant deterioration programs have allowed, and are projected to allow, the location of most new or modified facilities. Commission studies also indicate that new plant sitings and industrial expansion in the areas studied have not been significantly affected by the cost of air pollution requirements in the past (p 4.1–6).

It found that only two PSD permits had been denied (both of which were eventually approved).

The Commission believed that the effect of the Clean Air Act's requirements on national economic indicators had not been significant, and was not expected to be significant[10]. The Commission recommended that public participation be further provided for and encouraged. It made a number of recommendations for change while retaining the basic provisions of the Clean Air Act, with some simplification and relaxation.

The views of the regulators, of industrialists and of environmentalists were aired at great length before the Committee on Environment and Public Works of Senate (1981)[11]. The Committee (1982) emphasised the need for controls on acid rain at the end of its massive investigation:

1 except with respect to the issue of interstate pollution, the Act provides a sound legal framework for dealing with air pollution problems in the nation. As with any complex law, as more experience is gained in its implementation, modifications to improve its operation and clarify Congressional intent are desirable;
2 additional provisions are needed if the Act is to provide adequate means to address problems arising from air pollution that originates in one state and adversely affects health or welfare in another state (p 1).

The Committee reported a bill that contained a large number of clarifications, reduced the overlap between EPA and state and local air pollution controllers, simplified the PSD provisions and required sweeping reductions in emissions of sulphur dioxide.

No majority could be reached within Congress to enact these reforms and the Clean Air Act was merely reauthorised. The debate continued and the Committee on Environment and Public Works (1987) returned to the issue, reporting another bill which stressed the need for increased control over the precursors of acid rain and laid specific requirements on EPA to implement the Act, to prevent EPA leaving implementation to the states. It also proposed to outlaw bubbling and emissions trading for stationary sources in nonattainment areas because of what it regarded as widespread abuses. Generally, the Committee appears to have been exasperated at the lack of progress in achieving attainment: 'Over 100 million people are breathing air that does not meet the Federal health-based air quality standards' (p 2). There were dissenting voices about this, however, and Liroff (1986) has stated that trading 'has produced some noteworthy benefits, and further benefits will be lost if it is abandoned' (p 145). The need for reform of the Clean Air Act, not least to insert acid rain control provisions, and the need for better enforcement of its provisions are universally recognised but the political consensus necessary for revision had still not emerged at the time of writing.

Whatever the opinion held about the Clean Air Act, it is apparent from the size of the air quality control regions (which often extend for hundreds of miles) that, insofar as locational considerations are important in air pollution control decisions, these are regional rather than local in nature. The rules of the agencies, and the federal provisions relating to new stationary sources, take almost no account of the immediate neighbours of a new source. Thus, the PSD and nonattainment provisions are designed to protect regional, not local, air quality. It appears not to rest with the air pollution control agency to determine, for example, whether or not a new source should be located close to housing.

LAND USE CONTROL

Legal framework

Land use controls developed late in the history of the United States. Before the 1920s there were few attempts to control land use, other than the elimination of slaughter houses and gun powder storage areas from residential neighbourhoods (Sternleib et al, 1975). It was in the cities that it became apparent that regulations were needed to prevent one man's use of his land from depreciating the value of his neighbour's property. Rudimentary ordinances limiting the location of Chinese laundries appeared in San Francisco in the last years of the 19th century, and regulating building height and land use in Boston, New York and Los Angeles in the first decade of the 20th century (Delafons, 1969). In the next decade several cities passed local ordinances dividing real estate into districts which permitted some uses and excluded others. As Babcock (1966) put it: 'zoning was no more than a rational and comprehensive extension of public nuisance law' (p 4). In a landmark decision (Village of Euclid v Ambler Realty Corp[(12)]) the Supreme Court gave its blessing to this form of limited control on the use of land without payment of compensation in 1926.

The federal government has the authority to impose land use controls on state and local government directly. However, it has followed a permissive course and promulgated a Standard Zoning Enabling Act in 1924 which was readily followed in some form by most state legislatures, delegating zoning powers to the counties and municipalities. An enabling act to encourage the preparation of land use plans followed shortly after but was not so widely adopted by the states. During the period following World War II the techniques of zoning became more selective and flexible (and rather more closely related to land use planning criteria) while not greatly extending the discretionary powers of the planning agency. In the 1960s and 1970s, as well as concern about growth controls, there was increasing interest in various states in centralising some land use decisions. A federal land use planning act almost became law in the mid 1970s. Many of these initiatives have foundered and the smallest units of local government retain most control over land use decisions. Garner and Callies (1972) stated that:

> The complexity of the new techniques cannot obscure the fact that local zoning remains essentially what it was from the beginning – a process by which the residents of a *local* community examine what people propose to do with their land and decide whether or not they will permit it (emphasis in original).

The trend towards greater state involvement in land use planning (see later) cannot obscure the continuing truth of this observation. Each state has the power to enact legislation for the promotion of the public health, safety, morals and general welfare of its citizens. This so-called police power is the authority upon which state and local statutes regulating the use of land are based (Moss, 1977).

The courts have played a prominent role in the development of land use regulation in the United States. A series of court cases indicated that the concept of *taking* applied to government regulation of land. This limited the severity of land use controls which could be applied in the name of general welfare without requiring that the owner be compensated for the taking. Interpretations of what constitutes a taking have varied between the states but, as Bosselman et al (1973) pointed out, 'the popular fear of the taking clause is an even more serious problem than actual court decisions' (p 328). Further, they reported that 'the most recent court decisions ... strongly support land use regulations based on overall state or regional goals' (p 328). Callies, writing in 1980, found that 'the taking issue is being resolved in favour of environmental protection and land use control across the country'. It is perhaps symptomatic of the fluid land use regulatory context in the US that this may be less true today. Recent cases: San Diego Gas and Electric Co v City of San Diego[13], First English Evangelical Lutheran Church v County of Los Angeles[14] and Nollan v California Coastal Commission[15] suggest that regulatory takings may often occur and that monetary compensation will usually be the remedy. Although the taking issue is still not judicially resolved finally, these decisions have had a 'chilling effect' on local government land use regulation (Callies, 1985). However, provided regulations are based on properly prepared plans and conditions on the use of land substantially further the purpose for which they were enacted, they should not constitute a taking.

Even now, although all states permit the promulgation of local zoning ordinances, very few require all their constituent local governments to pass them. Some states even prohibit certain local governments from zoning. For example, Texas will not permit its counties to pass zoning ordinances. The result is that, in many states, only the urban agglomerations are zoned. Substantial areas of land, many of them in agricultural use, outside the boundaries of cities and villages remain wholly without land use controls.

Local jurisdiction land use controls

Land use planning
In theory, at least, land use regulation should follow comprehensive planning. The national standard City Planning Enabling Act 1926 contemplated the establishment of local planning commissions whose duty it was to adopt a *master plan* or comprehensive scheme for the physical development of the city. However, the plan was to be advisory to the elected representatives of the city. While many states have adopted legislation permitting the preparation of plans and their adoption by local legislative bodies, many do not require plans to be prepared and more do not demand conformance to these plans after adoption, though they can be enforced through sub-division controls on the land development process. Even where they do, the local elected body can overrule the finding of a non-elected planning commission that a proposed project is not in accordance with the plan – but this is rare. Thus the comprehensive planning which is undertaken is largely advisory in nature. In certain states, California, Hawaii and Oregon, for example, there is a requirement that zoning adheres to the goals and principles set out in the plan (Healy and Rosenberg, 1979). There is enormous variation in the role land use planning can play in a local community, ranging from guiding all public and private investment decisions to having no effect on such decisions (Clawson and Hall, 1973:168).

The broad process for adopting and updating comprehensive plans is similar for all jurisdictions. A planning staff, or a consultant, conducts studies and develops a draft plan for review by the public and by government agencies. A planning commission of nominated – not elected – members holds at least one public hearing and receives comments. The plan may then be revised and is forwarded to the elected representatives who then normally hold at least one public hearing before officially adopting the plan.

The comprehensive plan presents development proposals for public and private land within the planning area. It also specifies the allocations and locations of various land use categories including transportation and community facilities. The plans generally include sections on residential, commercial and industrial areas, recreation and public utilities. Notwithstanding the withdrawal of much federal funding, which appeared only late in their 50-year history, the use of comprehensive plans is growing.

Zoning

Together with subdivision control, local zoning is the principal means of controlling land use in the United States. In principle, at least, zoning should follow comprehensive planning (Callies, 1984). Most ordinances consist of a zoning map, district regulations, non-conforming use regulations and means of administration and enforcement. The zoning map for the local jurisdiction shows the various district classifications (industrial, commercial and residential, and their sub-categories) together with each exception or variance granted.

The most important part of any zoning ordinance is the allocation of land use districts and the establishment of permitted uses in each. These districts have their roots in the interests of the community in the use of land so as not to injure adjoining land. A modern zoning ordinance might have as many as 30 different classifications including light and medium manufacturing and heavy industry, single family residential and multi-family residential. It may specify population density in the various districts. Some zones are progressively inclusive so that, for example, in the industrial zone any type of use is allowed; in a commercial zone all but industrial uses are permissible, while the residential portions (the *highest use*) are the only exclusive areas.

Each zoning category contains an exhaustive list of permitted uses together with permitted accessory and special uses. A special use is generally small or unobtrusive and is only permitted in a district where it would not normally be allowed subject to meeting specified standards established by the local elected body and to certain articulated conditions (eg, a grocery shop in a residential area open for certain hours). Each district will also have a set of bulk regulations limiting, for example, the size of plot per unit, the permitted height of structures, minimum garden sizes and off-street parking requirements. In theory, the most intense use of the higher use district forms a buffer zone between the least intense higher use district and the *lower* use districts. Thus a row of apartment buildings in one zone might separate a single family district from a business or industrial zone. Once a district has been zoned, pre-existing non-conforming uses are permitted to continue in operation but not to be enlarged or to re-commence after a period of abandonment (Garner and Callies, 1972).

Once land is zoned in a particular category, a developer can develop a property as of right for the appropriate use, but subject to the conditions in the zoning classification, to subdivision laws and to other local regulations. He must also have a building permit, which must conform to the zoning classification, but this permit is ordinarily allowed as a matter of right if the compliance with zoning is clear (Clawson and Hall, 1973:174).

Because the Euclidean zoning system has proved arbitrary and clumsy, the concept of the *planned unit development* has been adopted to guide development of large tracts of land in a unified manner while leaving the developer freedom to be innovative. The developer presents a detailed land use and development plan to the local legislative body and pledges to build only in accordance with the site plan. The area concerned (which may be several hundred hectares in extent) is then re-zoned or granted a special use permit which overlays existing zoning regulations. Numerous uses within the overall area of the planned unit may be permitted by this means, reducing the uniformity of conventional zoning and increasing flexibility and the opportunity for negotiation between the developer and the local government.

The *floating zone* is a special use district which is defined by standards but which is not applied to any particular area until a developer asks for land to be rezoned in that category. For example, an authority might wish to establish a district for carefully planned multiple-family dwellings to serve a cross-section of income groups but leave the location to be decided by a developer. *Cluster zoning* provisions permit landowners who plan to subdivide their parcels and build on a smaller portion of the subdivision (ie, to cluster the houses) thus leaving the rest of the land as open space. Many jurisdictions use special overlay zones to protect particular areas (wetlands, rivers, historic areas, etc).

The local elected legislative body is responsible for enacting the zoning ordinance in its original form and for adopting amendments to the regulations and maps. Applications for amendment are filed with the local government, then often referred to the planning commission or zoning board which makes recommendations to the elected officials (with or without a hearing). The elected officials must give public notice and schedule a public hearing before deciding on an application. Local governments may also grant *variances*. A variance is special relief granted to an applicant from the strict requirements of the provisions of the zoning ordinance due to practical difficulty in meeting code requirements and where compliance would result in an unnecessary hardship for the applicant. Both special uses and variances are normally granted by a nominated board of appeals or zoning board of adjustment. Appeal is then sometimes to the local legislative body but more often directly to the state courts or (exceptionally) to the federal courts (Delafons, 1969; Garner and Callies, 1972). The appointed zoning administrator is generally the person responsible for seeing that the provisions of the ordinance are carried out. Increasingly, minor modifications to local land use controls are also delegated to this official. Zoning controls appear to be increasing in use and to be becoming more sophisticated and flexible.

Reflecting the fact that their origins lie partly in nuisance avoidance, most zoning controls are in effect performance standards, or proposed ways of distinguishing between different types of uses. Technical performance standards have been developed to deal with smoke, odour, dust, etc (Abt, 1977) (see Chapter 2).

Subdivision control
Most states have enacted statutes which require the preparation and governmental approval, often in both preliminary and final form, of a scaled and precise map (a *plat*) whenever a landowner proposes to subdivide a parcel of property into a number of smaller pieces. The purpose of this is to establish and enforce standards for public facilities (streets, sewers, water mains, etc) by depicting lots and blocks together with streets, alleys and utility easements. It is also a vehicle for requiring various dedications and donations (a form of *planning gain*). Most state enabling acts permit municipal corporations to elaborate on the platting requirements by setting minimum design standards, including the width of streets and pavements, the materials to be used in their construction and the placement of street lights, etc. The plat is usually submitted to the planning commission for review and then formally approved by the elected body, subject to the developer providing all the necessary infrastructure without public expense (Garner and Callies, 1972).

Subdivision control, or development codes as they are now often called, are often very similar to certain of the more flexible zoning controls, such as the planned unit development. Concurrent applications for zoning and subdivision permits can sometimes be made.

State land use controls

The local land use decision-making process in the United States has been roundly condemned because 'the criteria for decision-making are exclusively local, even when the interests affected are far more comprehensive' (Babcock, 1966:153). Pelham (1979) believed that:

> Thus, unfettered by broader areal policy considerations, local governments have systematically excluded locally undesirable development of state and regional benefit while actively promoting other types of development without regard for its adverse impact on state and regional values (p 1).

Almost every local government has sought to maximise its tax base and minimise its social problems (Bosselman and Callies, 1972) and there are thousands of local governments. Manifestations of the local control of land use have included urban sprawl, loss of farmland, degradation of the natural and physical environment, limitation of housing opportunity as well as fiscal inequity and discrimination (Healy and Rosenberg, 1979). There was often little public involvement in sensitive land use decisions (Rosenbaum, 1976).

It is small wonder that a movement was started in the 1960s to rationalise land use decisions: 'the quiet revolution in land use controls' (Bosselman and Callies, 1972). This sought to achieve a broader perspective in land use decision making by reclaiming for the states the regulatory power previously delegated to local governments. One manifestation of this movement was an attempt to pass national land use policy legislation to ensure that planning was implemented throughout the United States (Lyday, 1976).

The debate over national land use legislation began in 1970 and continued for more than five years. Though the bill was ultimately defeated very narrowly in the US House of Representatives, the policy debate played an important role in stimulating many states to assume greater responsibility for land use. Further, some of the objectives of the proposed legislation have been achieved by the passing of separate federal statutes dealing with issues having land use implications such as the Resource Recovery Act, the Coastal Zone Management Act, etc. Healy and Rosenberg (1979) argued that there was a role for state land use regulation, even if strong local controls existed, for areas of critical state concern, developments of regional impact, developments of regional benefit, unregulated areas and developments affecting or affected by major state investments (p 251).

As in the earlier zoning movement, the preparation of model legislation (by the American Law Institute, 1975) advanced the cause of reform. The Model Land Development Code 1975, with a provision for increased state participation in land use decision making, formed the basis for several state acts, including the Florida Environmental Land and Water Management Act 1972. There was clearly a trend towards considering land as a resource as well as a commodity in evidence in the 1970s.

Most state land use regulation falls into one or more of four categories:

1 direct state assumption of responsibility for regular zoning and/or subdivision control sometimes in areas otherwise unzoned;
2 statutory provisions designed to ensure that most new development should not cause any serious environmental damage, by means of (a) a state-level administrative review under general standards, and (b) the issuance of permits subject to conditions;
3 general requirements for regional-level review and permits, applying generally to specified geographic areas;
4 special requirements for protection of particular types of landscapes (Abt, 1977:334–335).

Thus the land use mechanisms adopted by various states include land use planning programmes, coastal zone management programmes, wetlands protection and management programmes, critical areas designation programmes (these areas usually contain scientific, historic or natural resources), power plant siting programmes, surface mining regulations, new towns legislation, environmental impact assessment processes and regulation of developments of regional impact (because of their size, environmental problems, traffic generation, etc). The adoption of such measures varies: for example, well over 40 states have a land use planning programme (some of which include state comprehensive planning or state enforcement of local comprehensive planning of some type). Rosenbaum (1976) has described the process of diffusion of many of these measures.

The typical pattern under a state permit programme for, say, its coastal zone is for an inventory and map of the area to be prepared, boundaries to be established (usually with public hearings), the adoption of permit regulations requiring a review of the proposed development and, in some cases, the adoption of more detailed land use restrictions. It is not unusual for states to have a dozen agencies with one kind of land use planning or regulatory responsibility or another, most set up as a consequence of receiving federal funds in return for the institution of a federally approved control system (Moss, 1977).

In the 1970s, there was a trend towards proposing more comprehensive land use plans and regulatory systems to control development over the whole area of a state. Hawaii, Vermont, Maine and Oregon have probably enacted the closest approximations to comprehensive state land use planning controls embracing formulation of policies, information requirements, co-ordination, participation, a central implementing agency, areas where development is precluded and an administrative appellate process (Healy and Rosenberg, 1979; DeGrove, 1984).

This trend was interrupted around 1975 as, with the exception of the California Coastal Act 1976, very few major state land use laws were passed over the next decade and concerted efforts have been made to repeal newly enacted land use legislation in some states. The reasons may include the ravages of the economic recession and the energy crises of the 1970s, the defeat of federal land use legislation, the lessening of growth pressures and political opposition to the centralisation of land regulatory power at the state level (Pelham, 1979). In addition, characteristic American attitudes: the spirit of individualism, the private property ethic, the prestige of the entrepreneur, the veneration of the market, the contempt for bureaucracy and the desire to limit government, militated against further reform of land use controls (Popper, 1981).

However, more recently certain states have begun to legislate in the land use control field again. Maryland and Virginia have recently passed coastal protection measures, New Jersey has passed protective legislation and Florida has refined and strengthened its control system (DeGrove and Stroud, 1987). Some have been disguising land use controls as regulations designed to protect groundwater. Ballot box zoning has even taken place in California where the national east and west coast concerns with traffic congestion, historic preservation and visual quality are perhaps most acute. The withdrawal of federal funding for infrastructure (especially roads and wastewater treatment facilities) is another factor causing states to question industrial development. As DeGrove (1984) has said:

continuing political support for policy and program initiatives ... are much broader than the narrow environmental concerns that often spurred their adoption ... But, the need for sensible tradeoffs in the economic versus environment arena must be faced by both environmental and economic development advocates (p 396).

Overall, 'it is clear that the movement to strengthen the state's role in growth management is winning new support in the 1980's' (DeGrove and Stroud, 1987).

Despite the various innovations, state controls have not been entirely successful. There has been a lack of co-ordination between state functions and none of the available state techniques (environmental plans, policies, co-ordinating councils, impact statement reviews, etc) have been sufficiently powerful to regulate land use effectively. Popper (1981) stated that the defects of state land use agencies turned out to be much the same as those of the zoning agencies they were intended to supplement. They did not have jurisdiction over large amounts of important development and were under-staffed and under-financed. Similarly, state efforts to impose some controls over the land use decisions of local governments, such as setting maximum processing times, have been fiercely resisted and have therefore had only limited success. As elsewhere, monitoring and enforcement are concerns that are gradually coming into their own in state land planning and regulation (DeGrove and Stroud, 1987).

Nevertheless, as Healy and Rosenberg (1979) pointed out:

> Over time, the movement toward greater state involvement in land use matters has had two persistent themes. First, it has increased the general level of consciousness of the environmental and other impacts of land development and has shown that stringent controls can indeed make a difference. Second, the experience of the states has shown that there are indeed non-local interests in how land is used (p 273).

Pelham (1979) reported that the effective implementation of the comprehensive planning approach required adequate funding and staffing and that penetrating the traditionally autonomous local regulatory system continued to be a difficult problem. The exponential increase in citizen participation since the 1960s was one reason why:

> Ironically, therefore, the chief legacy of the quiet revolution, which has become a code word for state recoupment of land-use regulatory power from local governments, may be the strengthening of local land-use controls (p 205).

Callies (1980) agreed that local government had re-emerged as a major force in the shaping of land use decisions and that local planning had regained impetus. He put this down to the proliferation of permits required (from numerous agencies at various levels of government) to undertake development, the federal incursion into the business of land use control (through the Coastal Zone Management Act, the National Environmental Policy Act, the Clean Air Act, etc) and the increased – and increasingly organised – citizen participation in land use decisions. He concluded:

> Indeed, it is a local, and not a state or regional, law which seems to have moved the country along the land use continuum to the preservation of the ... natural environment as a valid goal of land use regulation, provided private property is not altogether stripped of value.

This local planning impetus, and the state planning initiatives, should not, however, be seen as applying equally across the country. It does not apply in much of the middle of the country though states like Louisiana and Texas are now regretting their relaxed approach to the regulation of oil-related industry over the past decades.

Practice and reform

It is very difficult to gain a comprehensive picture of the numbers of professional planners employed on land use matters in the United States though the American Planning Association has estimated the number in the public sector at 30,000[16]. Many agencies employ none. There are also numerous lawyers employed on land use matters. Many planners are, of course, engaged in single sector planning with land use ramifications, rather than comprehensive land use planning. It is clear that the land use activities of many local governments have been severely affected by budget cuts over the last few years: both the quantity and, often, the quality of staff leave much to be desired.

There are, as has been discussed, also shortcomings in the land use control system. Delafons (1969) remarked that:

> It is hardly more than an historical accident, a singularly fortunate one, that there is any system of land-use controls available to American communities today ... The wise use of land and the orderly development of the community were little considered, beyond the elementary principle of separating grossly incompatible uses ... (p 106).

The zoning system is supposed to afford a specific set of controls. However, there is considerable discretion in the writing of zoning ordinances and they vary significantly from local government to local government. Zoning is used as an inadequate device to guide urban development, not usually to stop it or even to encourage it: it is essentially a mechanism for protecting private property. Thus, in a developed area, zoning will enjoy widespread support but in a developing area there will be little support from those who wish to develop when and where there is money to be made by doing so (Clawson and

Hall, 1973:170). There are, nevertheless, strong growth management movements in many high-growth areas.

Furthermore, the administration of zoning ordinances belies the seeming predictability of the system. There are few certainties, despite all the rules, even if there is an apparently strong presumption in favour of a particular development. A developer may be granted a permit subject to numerous conditions or be refused as a result of specially invoked rules (Babcock, 1966). It is not unknown for a jurisdiction to pass special legislation prohibiting a particular type of project anywhere in the locality, even if some permits have already been granted. Thus, notwithstanding the inability of a zoning system to prevent development in general, it is usually possible for a local legislative body to prevent a development it considers particularly undesirable by delaying or refusing minor permits until the developer abandons his application. Much land is zoned agricultural to allow discretion as to which types of development, if any, should be permitted and to keep property taxes low. Zoning is a highly political and legalistic activity and 'people do make a difference' in determining outcomes (Babcock and Siemon, 1985:256)

It is much more common for applications for zoning variances or exceptions to be approved regardless of their impact on adjoining areas (unless neighbourhood opposition is aroused) than for projects to be stopped. Thus, in the 1960s, about three quarters of applications for variances were granted. Some of these variances have been used to permit massive changes in land use and density, particularly in southern states. This abuse of the intentions of the variance has allowed developers to apply to nominated boards to modify zonings and increase the value of their land without any say by elected representatives in the granting of the permission. Unpredictability and ineffectiveness appear to be fostered by the absence of technical expertise and the presence of vested interests among the laymen who sit on zoning boards of appeals and by the numerical and technical inadequacy of planning staffs.

Business interests and political factors tend to dominate zoning and rezoning decisions taken by the elected councils: rezoning decisions are typically ad hoc, parochial, and based on narrow considerations. Again, about 75% of rezoning applications were granted in the 1960s. Babcock and Siemon (1985) saw 'parochialism, the NIMBY [not in my back yard] complex' as the fundamental problem in land use control and not, for example, corruption (p 260). They felt that the system was patently in need of reform, not least by reducing the 'incredible discretion left to the public decision makers' by accelerating administrative procedures and by removing 'demands for public gifts – exactions – in return for permission to develop' (pp 264–265).

Failure of land use regulation to shape growth led to attempts by several communities to slow or stop growth in their vicinity. Some municipalities decided that they would limit the annual rate of growth (eg, Petaluma, California). Such an approach to growth management, of course, tends to have the effect of merely redirecting growth elsewhere. These attempts to control growth have seldom had any air pollution control objectives. Several communities imposed moratoria on various phases of development (CEQ, 1974) and a number of new concepts of slow growth or timed development were successfully implemented (Godschalk et al, 1977). However, growth controls tend to be vehemently opposed by developers, to be highly sophisticated and hence to be difficult to administer. One celebrated example of the timed development approach, by the community of Ramapo, New York, was abandoned because of economic pressures and a slowing down of growth. Plans are now becoming flexible and many localities are buying more land to achieve land use objectives.

It is apparent that land use planning controls vary enormously across the United States. They have proved generally inadequate for the task of shaping growth and it might therefore be anticipated that they would be able to make but little contribution to controlling air pollution.

ENVIRONMENTAL IMPACT ASSESSMENT

In its fifth report, the Council on Environmental Quality (1974), reflecting the difficulties of using zoning to control US land use change effectively, stated:

> There is an increasing recognition that development proposals must be examined on an individual basis under a system of review that has both clearly defined standards and the flexibility to take into account changing community values and the special characteristics of each project (p 54).

Environmental impact assessment (EIA) is perhaps the best known technique for individual project appraisal. The EIA system was introduced in the United States on 1 January 1970, under the provisions of broad enabling legislation, the National Environmental Policy Act (NEPA). Since then the system has been substantially refined by judicial findings and by the issuance of new guidelines in the form of regulations by CEQ[17]. These regulations have been adapted and supplemented to meet their own needs by the various federal agencies responsible for the preparation of environmental impact assessments.

Interestingly, while most environmental legislation in the USA has become increasingly prescriptive, detailed and complex, NEPA was short, simple and comprehensive. NEPA nowhere provides for the termination of a major federal action regardless of the environmental consequences, but actions in the courts have stalled or stopped such projects if their consequences have not been properly documented (Callies, 1984:120–123).

The first step in the EIA procedure is the identification of the proposal leading to the action by the agency (construction of, or funding of, or permit granting for, a project). The agency itself, or the developer, will then undertake a preliminary scoping (a procedure intended to bring about agreement with the public as to the key environmental impacts requiring investigation) and environmental analysis to determine whether there is an obvious need for an environmental impact statement (EIS), whether the environmental impacts are clearly so minor as to permit a categorical exclusion from the EIA process (for which documentation is optional) or whether an environmental assessment should be prepared so that the impacts can be more clearly identified. Depending on the findings of the assessment, an EIS may be required or, as in the majority of cases, the agency may decide that none is necessary. If it is decided not to prepare an EIS, a finding of no significant impact (FONSI) must be written, summarising the reasons for this decision (CEQ, 1978).

When an EIS is required, scoping proper commences. The agreed issues are then addressed in the draft EIS. This is written by the agency though the developer will provide a great deal of the relevant information upon which it can be based if funding or permitting is involved. The draft describes the existing environment, explains what the proposed project is and analyses the effects of the project on the environment. It is these effects which constitute the substance of the draft EIS.

The draft EIS is sent to the Environmental Protection Agency (EPA) and is forwarded to all the relevant federal, state and local organisations and groups likely to wish to comment. Once the lead agency has received comments it is in a position to prepare the final EIS.

The final EIS describes the amended form of the proposed project, including any modifications that have been made since the draft EIS was published. The document normally contains quite extensive proposals for mitigation of impacts. A record of decision has also to be prepared, indicating the decision that has been made and the reasons for it. The final EIS should not normally be more than 150 pages long, according to CEQ regulations. Over 1000 EIS's were produced in many of the years since 1970 (Environmental Law Institute, 1981), but the number has dropped recently. In 1985, for example, there were 549 EIS's with a corresponding fall in the number of law suits to 89 in 1984 (CEQ, 1987). Indeed, the emphasis in EIA is moving toward mitigated FONSI in which negotiation takes place very early in the process and no EIS is produced.

There are somewhat inadequate provisions for monitoring the environmental impacts arising from an action and for ensuring that the various conditions or mitigation measures that have been included in the final proposal are implemented (Culhane et al, 1987). This may be done in the form of conditions appended to permits that have to be obtained from the lead agency or in the form of conditions attached to grants that are made by the agency. If the agency itself is carrying through the measures there is usually a system of inspection to ensure that the project is actually constructed as described in the final EIS.

It is normal for the EIS to address the procedural requirements of NEPA, as refined in the agency guidelines, and to rely on scoping for the identification of issues, rather than to use any 'comprehensive EIA methodology'[18]. Widespread use is, however, made of specialised technical methods for assessing particular impacts (eg, air pollution modelling). The trend is to make greater use of the information generated for other purposes (for example, the granting of an air pollution permit) in preparing the EIS and to combine the granting of permits to reduce the number of steps an applicant must make.

Overall, it would appear that the methods and procedures used in the environmental impact assessment process have improved the quality of environmental decision making in the United States. There have been costs, of course. In particular, delay and the expenditure of manpower and financial resources have resulted from the EIA process. Many environmental documents have been characterised as being little more than 'boiler place' (excessively verbose). On the whole, however, it would appear that this boiler plate mentality is declining and that the utility of EIA is being more widely recognised, despite its categorisation as a 'standard administrative reform measure' by some commentators (Fairfax and Ingram, 1981). Moreover, some of the net costs have been positive, with real savings resulting from careful prior consideration of environmental impacts and delays being minimised (eg, Canter, 1983).

There is no doubt that the EIA process is biting. Several projects have been aborted as a result of the adverse impacts revealed in preparing an EIS and it appears that a majority of projects are modified as a result of the assessed impacts. This mitigation of impacts appears to be 'where the action is' and is widely cited as one of the main justifications of the process. There has been substantial EIA litigation, initially by environmental groups but increasingly by industry which has begun to view the EIA system favourably (Liroff, 1981). The volume of litigation is now beginning to decline as many issues have been clarified. To a large extent, EIA has been assimilated into the federal decision-making process and is meeting many of the objectives of its proponents. Its current lack of notoriety may well be a measure of its success in internalising the consideration of environmental quality in federal agencies.

Despite the generally accepted improvement in the quality of EIS's over recent years (the quality of coverage of air pollution impacts in EIS's appears generally to have been adequate), there is scope for further amelioration in their analytical content (CEQ, 1982b) and for closer adherence to the spirit, rather than the letter, of NEPA (Indiana University, 1983). Environmental impact assessment has been mainly confined to projects and probably owes much of its success to the general weakness of the US land use planning system. Reilly (1974) has stated that the impact statement process 'reflects a more realistic understanding of the way major development is sited. No one any longer expects comprehensive plans to detail precisely the nature and location of new development' (p 350).

A number of states have enacted environmental impact assessment legislation, as have some counties and cities. Eight states require EIS's only for projects proposed within specific areas, four require them for actions undertaken by state agencies or using state funds, and seven require them for these categories together with actions requiring state permits. The requirements of another four states apply to all these types of actions plus a number of actions taken by local agencies, and three states, including California, have a comprehensive system covering local government and private activities as well as those of the state itself (Hart and Enk, 1980). The various legal requirements differ from the federal system and most have proved weak and ineffective, the comprehensive systems being among the exceptions.

This extension to state and local actions from federal actions has meant that EIA can effectively complement the land use planning process, though it should not be considered either as an additional, unnecessary burden or as a substitute for effective policy making in land use planning (Pearlman, 1977). However, only a few cities have adopted EIA procedures. Thus, at local level at least, the control of air pollution from stationary sources has to rely on more traditional procedures than EIA.

NOTES

1 See, for example, Council on Environmental Quality (1970) where it is stated that 'Misuse of the land is now one of the most serious and difficult challenges to environmental quality ...' (p 165).
2 The *Clean Air Act 1977* s 169A(b)(2) A requires sources which may reasonably be anticipated to cause or contribute to any impairment of visibility to install the *best available retrofit technology.*
3 Notwithstanding the broad coverage of the Clean Air Act, it contains no provisions to counteract the growing problems of acid rain.
4 See, for an interesting account of the Senate and House of Representatives compromises reached in passing the 1977 amendments, Domenici (1979).
5 Code of Federal Regulations Chapter 40, Part 60 (40 *CFR* 60).
6 40 *CFR* 60.
7 *Clean Air Act 1977* Part C and 45 *CFR* 52676.
8 Quarles, 1979. The Act contains provisions for pollutants other than suspended particulates and sulphur dioxide to be made subject to PSD requirements at a later date.
9 *Clean Air Act 1977* Part D.
10 Using the Commission's ranges of figures, annual costs are $4,000–16,600 M and benefits $4,600–51,200 M. The conclusion that net benefits accrue from the operation of the Clean Air Act, while not inescapable, is strongly supportable.

11 There were some six reports in all, several running to well over 800 pages. They constitute a goldmine of opinions on the effectiveness and shortcomings of the Clean Air Act, as they contain the views of virtually every acknowledged expert and interest in the USA.
12 *Village of Euclid v Ambler Co* 272 US 365 (1926).
13 *San Diego Gas and Electric Co v City of San Diego* 450 US 621 (1981).
14 107 S.Ct. 2378 (1987).
15 107 S.Ct. 3141 (1987).
16 Cobb, R L, American Planning Association. (1985). Personal communication.
17 The various guidelines (eg, CEQ, 1978) are published in the annual reports of the Council on Environmental Quality.
18 For a review of methodologies (matrices, networks, etc) see Canter (1977).

4
Implementation of preventive controls in the United States

It would, perhaps, appear from Chapter 3 that air pollution controls, land use controls and environmental impact assessment have been used largely independently of each other in the USA. These controls all impinge on the siting of new stationary sources of air pollution and the ways in which they have been employed in practice are recounted in this chapter. The first section of the chapter contains a discussion of the relationship between land use and air pollution controls. The remainder of the chapter consists of eight case studies, in seven states, of the implementation of US land use controls and air pollution controls over new stationary sources, ranging from a creosote storage depot to an oil refinery.

RELATIONSHIP BETWEEN LAND USE AND AIR POLLUTION CONTROLS

The potential importance of the relationship between land use planning and air pollution control has been recognised in some quarters in the United States for many years. Thus, the Council on Environmental Quality (1970) recommended that:

> Land use planning and control should be used by state, local and regional agencies as a method of minimising air pollution. Large industries and power generating facilities should be located in places where their adverse effect on the air is minimal (p 90).

Train (1976) believed that the 'formulation of land use policy has become indistinguishable from the formulation of environmental policy'.

There are two distinct aspects of the relationship between land use and air pollution controls: the effect of air pollution controls on land use and the effect of land use controls on air pollution. It has been stressed repeatedly that the Clean Air Act would force land use planning decisions to be made by air pollution control agencies[1]. As Train (1976) stated, 'the act, by demanding that ambient standards be met everywhere, necessarily demands land use controls'. In particular, the prevention of significant deterioration and nonattainment programmes, neither of which was anticipated in the 1970 Clean Air Act, have been considered to have profound land use implications.

Initially, it was believed that the designation of areas, often the longer-established industrial regions, as nonattainment for various pollutants might force industry to look elsewhere, reinforcing a trend towards growth in the south and west which was already strong. Similarly, it was thought that the PSD regulations might lead to urban growth of a very dispersed kind as increments were consumed:

> Any PSD policy, stringently enforced, will not only affect regional growth patterns and the siting of major industrial facilities, but will place states in the role of regulating land use through their air quality responsibilities (Manners and Rudzitis, 1982).

Indeed, it has been suggested that the PSD programme was itself developed to forestall the rush of industry from the north and east of the USA[2]. However, as demonstrated in Chapter 3, the widespread concern that the Clean Air Act has been responsible for preventing or restricting growth appears to have little basis in fact. Remarkably few applications to develop in nonattainment areas, or in areas where PSD applies, have been turned down. Thus, of the eight states visited as part of this study, only Californian air pollution controllers could aver that offsets for major developments had ever proved unobtainable[3]. In no case had PSD permits not been forthcoming.

While fears about the de facto implications of the Clean Air Act for land use have proved unfounded to date, the 1970 Act quite specifically stated that land use controls were to be one of the tools for implementing national ambient air quality standards, together with transportation controls[4]. Indirect source review was another land use related element. However, the implementation of land use controls, notwithstanding a great deal of research financed by the Environmental Protection Agency, became an extremely political issue as a result of an attempt to introduce parking controls in nonattainment areas through the use of indirect source regulations (Moss, 1977:40–67). These required a preconstruction review of shopping centres, stadia and other indirect sources to ensure that health related standards would not be violated because of increased vehicular traffic. This led to a Congressional backlash against federal intervention in local land use affairs and spending allocations for the programme were discontinued. EPA was forced to retreat by withdrawing the relevant regulations. As Manners and Rudzitis (1982) said:

> The transportation controls (including compulsory car pooling and restrictions on gasoline sales proposed) [in 1973] for the nation's most polluted cities reached the most profound implication of the Clean Air Act for the average citizen – the impact of the law on his relation to his own automobile.

Air quality alone was obviously not usually a sufficiently strong incentive for local governments to undertake urban planning, especially as land use responsibility is for the most part an activity of local government, while air pollution authority, particularly for stationary sources, is concentrated at the state level. The land 'use controls' requirement was deleted in the 1977 Clean Air Act. The other land use-related requirements remain and general land use controls are not precluded. It is still, of course, perfectly possible for a state or local government to administer, for example, an emission density zoning programme (Roberts et al, 1975).

Turning to the effect of land use controls on air pollution, there is little evidence of the various planning techniques being utilised in environmental management (Kaiser et al, 1974). Practice has lagged some way behind theory and land use planning in the United States has frequently been ineffective, not only in preventing land misuse but in protecting environmental quality. This has probably been because, for the most part, protection of environmental quality has not been fundamental to the planning process. Further, single purpose planning has resulted in a lack of co-ordination and, as shown in Chapter 3, the land-based tax structure and absence of centralised planning and implementation authorities have resulted in land use plans being ignored.

It is apparent, despite the various initiatives to improve the situation, that the complexity of environmental problems, the absence of proven environmental strategies, the shortage of financial resources, the lack of staff expertise and the fragmentation of government continue to frustrate the efforts of local governments to come to grips with environmental management problems including air pollution control (Magazine, 1977).

By the mid-1970s very few local governments and councils of government had conducted air quality planning, or even included air quality as an element of a comprehensive land use plan (Abt, 1977:121). The situation has not improved appreciably since, not least because of the decline of the councils of government and the reductions in budgets for planning. Where air pollution policies are included, they are often little more than pious hopes, such as expressions of the desirability of reducing car travel, encouraging car pools and staggering working hours. Similarly, while there are exceptions, few zoning ordinances make any reference to air pollution control. In the circumstances, it is hardly surprising that land use permits have sometimes been granted to the developers of new or modified stationary sources of air pollution with scant regard to the pollution likely to ensue.

The seven states chosen for study were selected to represent varying strengths of land use and other environmental controls using measures calculated by the Conservation Foundation. Duerkson (1983) used measures derived from 23 different indicators of environmental control. The measures had a possible range of 0–63 and the 50 states ranged from 10 (Alabama) to 47 (Minnesota). The scores of the case study states (with their ranks in brackets) were Louisiana, 21 (43); Texas, 22 (39); North Carolina, 25 (29);

Florida, 31 (17); Maryland, 37 (7); Oregon, 42 (5); and California, 46 (2). Table 4 presents Duerkson's seven indicator values of most relevance to this study for the selected states. Once these had been determined, the case studies were chosen for their intrinsic interest (Chapter 1). Table 5 summarises some of their main features.

THE LOUISIANA CREOSOTE STORAGE FACILITY

Palmer Barge Line Inc sought to construct a storage tank on a coastal 0.3 ha site in St Tammany Parish so that creosote could be transferred from barge to truck. A creosote manufacturing plant in the parish had previously burnt down, causing serious water pollution. The nearest dwellings to the site, occupied predominantly by well-to-do residents, were some 300 metres away and the surrounding land was in low intensity industrial (mostly storage) use. Palmer felt that it would be a straightforward matter to obtain the necessary land use permit from the parish. It did not expect to have to obtain air pollution or coastal management permits from the state.

The company wrote to the parish and to both the Coastal Management

TABLE 4   Selected environmental and land use control indicators for the case study states

Indicators	State EIS process	Priority to environmental protection	$ for environmental quality control	Comprehensive land use planning	Environment is stated goal	Specific state land use laws	Aesthetics and zoning	Total
Possible score range	0–4	0–4	0–6	0–4	0–2	0–6	0–2	0–28
LA	0	1	0	0	2	3	1	7
TX	2	2	0	1	0	4	0	9
NC	4	1	2	1	0	2	0	10
FL	0	3	0	4	0	5	2	14
MD	4	1	4	1	0	6	0	16
OR	0	2	4	3	2	5	2	18
CA	4	4	3	4	2	4	2	23

Source: Duerkson (1983)

72 *Planning Pollution Prevention*

TABLE 5 Some characteristics of the US case study siting processes

	Air Pollution Control Agency				Land Use Planning Agency					Other procedures		Implementation				Use of the courts
	Number of agencies	Prior objections	Accept/ refuse	Appeal (Acc/Ref)	Number of agencies	Prior objections	Accept/ refuse	Appeal (Acc/Ref)	Air pollution control conditions	Prior objections	Accept/ refuse	Construction	Subsequent objections	APCA enforcement	LUPA enforcement	
Louisiana	1	✓	✓		2	✓	×	×		✓	—					✓
					✓	×	×									
Texas	1	✓	✓	—	1	✓	×	×				✓				✓
						✓	✓	✓								
North Carolina	2	✓	✓		3	✓	×;✓	✓		✓	—				✓	
						✓										
Florida	3	✓	✓		2	✓	✓		✓	✓	✓	✓	✓	✓		
Maryland Providence	1				1											
Childs	1	✓	✓		1	✓	×	×		✓	✓	✓	✓	✓	✓	✓
Oregon	2	✓	✓		1	✓	✓	✓	✓	✓	×					
California: Chevron	2	✓	✓	✓	1	✓	✓	✓		✓		✓		✓		✓
		✓	✓													
California: Dow	1	✓	×	—	2	✓	✓	—		✓	—					✓
					✓	—										

Section and the Air Quality Division of the Louisiana Department of Natural Resources in May 1981. Within a few days the parish began to receive letters of protest about the risks of flooding, spillage and recurrence of the problems previously experienced with creosote. The parish zoning commission hearing was addressed by Palmer's lawyer, who was able to state that the parish's was the only permit needed. Several objectors spoke out strongly and the commission denied the permit.

The company appealed against this zoning decision and mounted a more impressive case before the parish council, largely addressing the problem of accidental spillage. The plant was expected to produce $200,000 per annum in local revenue from sales tax. Again, objections were raised and, again, the appeal was rejected. Palmer appealed to the courts but judgement was postponed on a procedural point: the need for written assurance that no state permits were necessary. The council of St Tammany Parish then passed an ordinance specifically aimed to prevent the construction of premises for storing or manufacturing creosote anywhere in the parish.

The air quality division of the Louisiana Department of Natural Resources wrote to Palmer stating that no permit would be needed on 27 May 1981. However, the officials later had second thoughts and asked the company to submit an application in July. This was prepared by consultants. Further information was requested and the officials decided that the decision to grant the permit (which was never in doubt) should be taken by the state Environmental Control Commission. This body held the decision over pending the sister Louisiana Coastal Commission's deliberations.

The Palmer letter to the state department's coastal management section indicated that all the site was more than five feet above sea level. As sites above this height are exempt from many coastal management controls, an oral indication was given to the company's attorney that no permit would be necessary. Letters of objection soon began to arrive in the coastal management section. In June more information was demanded of the company and a St Tammany parish councillor asked that no construction be permitted. The coastal management section officials began to change their minds as they learned more about the facility (including the fact that some of the site might be below the five foot level) and as opposition became more marked. The intervention of a state representative objecting to the use was important in this alteration of attitude. In July, without waiting for receipt of the information requested, the section decided that a permit would, after all, be necessary.

The public hearing into the coastal management permit application attracted some 30 speakers, the vast majority of whom were against the development, despite the tax revenues the facility would generate. The officials, after weighing the evidence, refused the permit.

Palmer Barge Line appealed against this decision and a hearing before the Coastal Commission took place in January 1982 with a hearing officer appointed to conduct the proceedings. The Palmer case was presented by several experts and was far more convincing than its earlier and less technical arguments to the St Tammany zoning commission. The objectors called the state representative, environmental scientists and a coastal management official. Despite being impressed with the Palmer case, the commission voted 12 to 3 to reject the appeal. A strong tradition of local autonomy prevailed, the commission declining to sanction a development so strongly disliked locally and opposed by the local commissioners.

Although confident of the strength of its case, Palmer had already spent over $50,000 in presenting it and there was every expectation of more expenditure to come. (The state had by now decided that a water pollution permit would also be required.) Accordingly, the company decided to cut its losses and withdraw the proposal. Palmer then arranged to establish the storage facility in the state of Mississippi but, though it was welcomed there, the company never constructed its plant because of the downturn in the economic climate.

The history of the creosote storage plant demonstrates the high level of discretion available at state and local level to refuse and delay projects, even where land use controls are weak. The refusal of zoning approval, the passing of the local ordinance, the change of opinion about the coastal use permit, the refusal to allow the appeal and the referral of the air pollution permit to the commission after stating that no permit would be needed, all appear to have been motivated by strong and influential opposition to a minor industrial development which happened to involve a locally controversial substance. While air and water pollution permits would have been forthcoming, the outcome of appeals to the courts on the zoning permit and coastal management permits would have been much less certain. The vacillation of the state officials appears to have partly been an indication of staffing weaknesses and partly an indication of the political (and parochial) nature of the siting process. The benefits of the project (which was proposed by an out-of-state developer), in local revenue and employment terms, were not great.

THE TEXAS ASPHALT PLANT

In 1982 Petroplex Land and Development Co Inc sought to erect an asphalt batching plant, which had previously been used in Alabama, just outside the city limits of Midland, Texas. Midland County had no zoning or building controls and no land use permit therefore appeared to be necessary. In July Petroplex applied for a standard exemption from the Texas Air Control Board's permit requirements on the grounds that the plant would be located more than half a mile from the nearest residence and that emissions would be below specified levels. The exemption was granted on the condition that the plant continued to comply with these requirements.

There was no indication that this decision would prove controversial and construction of the plant was duly commenced in the centre of a one mile by one quarter of a mile plot – just a few feet over half a mile from the nearest house. This was part of the Skyview Addition, a newly constructed group of expensive properties, also just beyond the Midland City boundary. Protests soon poured into the board from the Addition residents who demanded a public hearing to review the exemption decision.

The City of Midland was in the process of annexing land to expand its area at the time. While the residents of the Skyview Addition did not want their land to be annexed (they already had water and sewerage), they encouraged the city to annex the land on which the asphalt plant was being constructed, and to zone it (in accordance with the city ordinances) for 'agricultural estate' (low density residential) use but not for commercial use. The city acceded to the residents' wishes and annexed the Petroplex site while construction of the plant continued, but excluded the Skyview Addition.

The city forced Petroplex to apply for a building permit and temporary use permit now that the site fell within Midland's jurisdiction, though it could have chosen to regard asphalt making as a non-conforming use. Petroplex reluctantly applied and, in the meanwhile, the city passed a resolution strengthening the existing ordinances and requiring cessation of building until the necessary permits had been obtained. At the permit hearing at the city council meeting (following much publicity), the lawyer for the residents argued that asphalt manufacture was inappropriate in an area zoned agricultural estate and so close to existing housing. Numerous objectors spoke and Midland denied the use permit.

Petroplex continued construction, however, and in early December ran the plant at maximum output, producing a 'volcano-like' emission of dark smoke. The city decided that legal action to enforce its resolution was necessary and sought an injunction to force the cessation of operations. The company voluntarily halted manufacture before Christmas.

The level of controversy and complaint over the Texas Air Control Board's permit exemption was such that the first hearing relating to an exemption in that state took place in January 1983. (Petroplex had been warned by the board in the autumn that they proceeded at their own risk, once the hearing had been conceded.) The residents, Petroplex and the board were all represented by lawyers at the three day hearing. The residents argued that polluting industry should not be permitted to locate so close to housing. This argument was not admitted because 'proper land use is not a criterion for an exemption or construction permit' but was a matter for the local zoning authorities. The company argued that emissions would be minimal and the board, presenting a separate case, quoted the results of modelling exercises to demonstrate that concentrations would be very low and that it had been right to grant the exemption. In the meanwhile, Midland County had imposed weight limits on the roads from the site running closest to Skyview Addition to divert heavy vehicles serving the plant from several local roads.

The opponents of the plant had spent some $15–20,000 fighting Petroplex and had won substantial local television, radio and press publicity. They expected their case to be rejected by the hearing officer and thought that Petroplex would be granted a special air control permit exemption with further conditions to limit emissions (for example, surfacing of roads).

However, later in 1983 the previous owner of the asphalt plant removed it from the site without its operating again. Petroplex ceased to exist before the Texas Air Control Board could make a decision on the exemption permit or before the Midland City case went to court. The ability of a concerted opposition to influence a local government, in which they were not resident, to use its discretionary land use powers was notable. The annexation, the passing of the ordinance, the demanding of the use permit, the refusal of the permit and the taking of court action eventually led to the removal of the air pollution source, something that would have been almost impossible using air pollution control powers. The likelihood of further development in the vicinity of the plant, which had few employment or local revenue advantages, must have influenced the city council which has constantly annexed land over the years.

Petroplex took the position throughout that it would meet all its legal obligations (though it disputed the legality of the City of Midland land use resolution) but would not give an inch to the objectors, despite the ease with which it could have moved the plant. It is ironic that such controversy should have taken place over a permit exemption rather than a construction permit and that locating the plant in the first instance at the far end of the site, about a mile from the nearest houses, would probably have satisfied the residents.

THE NORTH CAROLINA OIL REFINERY

The Brunswick Energy Company (BECO) sought to build an oil refinery in a sparsely populated area on the banks of the Cape Fear River in Brunswick County, North Carolina[5]. The refinery was to be sited some distance from the river and surrounded by a buffer zone of existing woodland. As this was a coastal county, the state's coastal area management act applied and a land use plan had been prepared for the county[6]. No zoning provisions were in existence but, since the development was not in accord with the land use plan and was in an 'area of environmental concern' an amendment would have to be prepared by the county and approved by the state Coastal Resources Commission before the refinery could be constructed. An oil refinery, a major source of air pollution, would also require a prevention of significant deterioration permit from the Environmental Protection Agency and state construction and operation permits. In addition, constructors of oil refineries in North

Carolina must obtain a specific facility permit. Further, because a Corps of Engineers dredge and fill permit would be necessary, an environmental impact statement would have to be prepared. Water pollution, waste disposal and other permits would also be required, making about a dozen in all.

BECO appointed a project engineer who took the approach of willingly endeavouring to satisfy every environmental requirement. He commissioned a firm of consultants to prepare the material required to obtain the air pollution permits and to undertake other environmental permit application work. Local residents and local and state officials were taken to visit other refineries.

The officials of the North Carolina Department of Natural Resources and Community Development endeavoured to take a neutral stance on the various permits required for this, one of the largest developments ever proposed in North Carolina. The nominated head of the department, however, appeared to have prejudged the issue by publicly assuming that it would be built. The department appointed co-ordinators to try to ensure that all the various permits were handled with the minimum of duplication, and on a critical path. It also set up a citizens' liaison committee to ensure that discussion took place between BECO and potential opponents of the refinery.

A local organisation, Carolina Coastal Crossroads, was set up to oppose the refinery, choosing its title to indicate the fundamental change in coastal environment and lifestyle its members felt the refinery would cause. They had the backing of a wealthy property owner, who made a film for them and encouraged them to use the Washington law firm he retained to help prepare arguments. They not only organised petitions and appeared at public hearings but gave regular media interviews. (*The Wilmington Star* newspapers opposed the project.) They also used car bumper stickers and paid for outdoor advertisements. The majority of the local population was, however, in favour of the project.

Commencing in 1979, comprehensive monitoring of air pollution levels, both close to and some 35 miles from the site, was undertaken and a subsequent modelling simulation was carried out by the consultants. The submission of an application for the various air pollution permits consisted of two volumes of data and argument. The Department of Natural Resources and Community Development negotiated onerous air pollution controls which would have involved the utilisation of only a small PSD increment and, notwithstanding strong public protests, issued a preliminary notification of approval some 10 months after receiving the application and announced that a public hearing would be held before final approval was granted by the state. BECO felt unable to accept one or two of the conditions, especially one requiring exhaustive and expensive epidemiological studies. The air pollution control officials felt they had used the available federal and state regulations to reduce air pollution from the refinery to the minimum feasible.

BECO produced a voluminous environmental report as a basis for the

78 *Planning Pollution Prevention*

Corps of Engineers' environmental impact statement. This served also to provide the information necessary for various other permits and revealed a number of matters requiring design modifications.

The elected representatives of Brunswick County were much in favour of the development, which would employ around 400 people in operation (2–3000 during construction) and yield a high tax revenue. They accordingly approved a county land use plan update which redesignated the site (and other large areas) from 'conservation' to 'industry' use. This approval followed a noisy public hearing at which the justification of the oil refinery became the main issue. The major last minute alterations to this plan betrayed an absence of concern for planning principles and the weakness of the local land use planning system. The Coastal Resources Commission, however, rejected the land use plan on procedural grounds, as the county had not allowed adequate public notice of the changes to the plan before the hearing. A further county hearing was held and, despite official misgivings and the expression of objections from neighbouring New Hanover County and the nearby town of Wrightsville Beach at a state hearing, the commission approved the revised plan in March 1981.

Shortly afterwards, in May 1981, BECO decided to withdraw its application because of declining demand for oil products. By then the estimated cost of the refinery had risen from about $400M to $1,000M; BECO had spent two and a half years and about $3M on the permitting process and was six months behind its anticipated schedule.

This case history illustrates the large number of permits that can be involved in building a major pollution source. While the granting of these is normally not at issue, delays can arise, as in the air pollution permit process. This project brought substantial local benefits and was welcomed by the local government to which they accrued. The refinery was carefully located away from the river and designed to make use of the existing woodland as a buffer zone to reduce air pollution. Even so, and despite strenuous public relations initiatives, the developer encountered well-organised opposition, though BECO managed to avoid courtroom entanglements. Given its genuine willingness to meet environmental requirements by employing the appropriate technology, the company would probably have received its remaining permits within relatively short order, despite the effectiveness of the opposition from Coastal Carolina Crossroads. Ironically, the decision involving the greatest use of discretionary powers, the amendment of the county land use plan, was the only one actually made when BECO withdrew.

THE FLORIDA RESOURCES RECOVERY FACILITY

Metropolitan Dade County had long had a severe solid waste disposal problem and turned for a solution to the resources recovery concept in which materials are recovered and electricity generated. The sale of bonds to finance the project was agreed in 1974 and a private company, Resources Recovery (Dade County) Inc (RRDC) was set up to manage the project and operate the facility. It was decided that it would probably be easier for Dade County to apply for all the relevant construction permits as owner of the facility.

A public hearing into Dade County's zoning application[7] to construct the facility to the west of Miami, over a mile away from existing housing but close to an existing landfill site notorious for polluting ground water, took place in 1975. No member of the public wrote to object or appeared in person to make representations, presumably because nobody lived near enough to the site to be concerned. Dade County's planning department recommended approval and altered the comprehensive plan designation from agricultural and open land to industrial land use. The net environmental benefits were expected to be substantial, as the existing landfill would be closed. Accordingly, in a markedly non-controversial decision, the Board of County Commissioners gave its permission for the county development to proceed.

Florida has a power plant siting law which applied to this plant as electricity was to be produced. A power plant certificate – a 'one-stop permit' – subsumes most other state and local permits but, because of the length of time usually involved in certification and the necessity to obtain funding approval and to commence detailed design very quickly, Dade County decided to obtain the state and local permits first. Permission would allow construction to start. Accordingly, Dade County applied for state air pollution construction permits in 1976 and these were granted in 1977, without any objection, on the assumption that the plant would operate six days per week. Particulate emissions were not to exceed 0.08 gr/ft^3. Application was also made for a federal PSD permit which was granted, six months later, in 1978. Dade County, RRDC and the state and federal air pollution control agencies were anxious that the calculated hydrocarbon emissions would not cause the incinerator to be classified as a major source for hydrocarbons, as Greater Miami is an ozone non-attainment area and the federal Clean Air Act offset provisions would apply.

Application for the power plant siting certificate was made in 1977. The consultants' documentation, which was similar to an environmental impact statement and took a considerable period of time to prepare, covered air pollution among many other topics. The state's power plant siting office reported that Dade County's sulphur dioxide standard (8.6 µg/m^3) might be violated but the state (60 µg/m^3) and federal standards for this gas and for particulates would not be exceeded. Following public hearings in 1977, at which there was again no objection from the public, a certificate was granted in 1978, five months after application had been made. This was subject to numerous air pollution control and other conditions to limit environmental problems. Because RRDC now decided that seven day per week operation of the plant and some design changes were desirable, it became necessary to obtain a new state construction permit. This was applied for by RRDC and was rapidly granted.

There then arose a dispute between RRDC and Dade County about payment for construction and the price to be paid per ton of refuse treated. Construction of the plant was completed in 1981 and, although there was considerable doubt about whether any further permits were necessary (since a power plant siting certificate had been granted), RRDC applied for and received one of the several state air pollution operating permits required. The state's lawyers determined that, since the certificate had been issued to Dade County, RRDC would still have to obtain the various relevant permits, as operator.

In 1982 Dade County advised RRDC that it too required air pollution operation permits, renewable annually at a cost of over $2,000. State and county solid waste facility operating permits were also needed. RRDC had either declined to apply for these or not been granted them in 1983 when the state reversed its earlier position and ruled that state and county solid waste and air pollution permits were, after all, subsumed by the power plant siting procedure. The Corps of Engineers belatedly decided that a federal dredge and fill permit would be needed. This was rapidly granted.

Both the county and the state were becoming concerned about air pollution from the plant. Various letters were sent to RRDC from the county and the state (which the county constantly chivied to enforce its construction permit) about odours and visible emissions. These were always countered by RRDC stating that these problems were only sporadic and typical of most industries. Dade County was very disappointed by the plant's performance which the plethora of regulation seemed to have done little to mitigate. Numerous complaints from residents to the west of the plant now began to be received and these were instrumental in persuading Dade County to terminate its contract with RRDC. A new management company has subsequently improved the plant's pollution performance and achieved a reduction in the level of complaints.

The regulatory confusion that can arise in the United States is illustrated by this case. The one-stop power plant siting process took only five months to complete and, in the end, was ruled to subsume most of the other overlapping state and local permits. The power plant siting permit was notable for the numerous air pollution control conditions attached to it. However, neither these nor the state and local air pollution permit conditions proved particularly helpful when the resources recovery facility started to cause pollution problems. Indeed, it took a change of management, and therefore of attitudes, not the enforcement of land use or air pollution control powers, to achieve any real improvement in pollution levels.

THE MARYLAND SOLVENT RECYLCLING PLANT

In 1961 the Galaxy Chemicals Company opened a small solvent recyling plant on the site of a former paper mill at Providence, Cecil County, Maryland, in the steep-sided, sparsely populated valley of the Little Elk River. This was before the county's 1962 zoning ordinance was passed, in which the site was designated for industrial use. Numerous complaints ensued, many of them from a local doctor resident in the valley who claimed that he had detected benzene, toluene and other carcinogens in the air and in the river and that the level of cancer in the local population was much higher than expected. In 1970 the Maryland Department of Health and Mental Hygiene was granted an injunction to prevent the company from emitting odours and the plant was closed down for extensive refurbishment in 1971.

The doctor moved away but continued his campaign and attracted national publicity, through the *Washington Post*, for his findings about 'Cancer Valley'. A state investigation took place in 1974 but, though the presence of odours and elevated cancer levels were acknowledged, it was never possible to prove that these were due to the Galaxy plant. Galaxy Chemicals was declared bankrupt in 1975, but recommenced operations at Providence as Spectron Inc. Some 20 legal cases relating to pollution involving the company had taken place between 1969 and 1975, many of which it lost. The owner of the company, a chemical engineer, trained as a lawyer to lead his own cases.

Cecil County politicians and officials were obviously perturbed by the pollution incidents and by the gradual expansion of the plant. When a county-wide rezoning took place in 1979, it was decided (not without considerable controversy) to rezone the area concerned as 'agricultural', thus making the plant a non-conforming use and rendering legal expansion virtually impossible. Nevertheless, expansion continued amid much bad feeling and with considerable legal activity which began to affect Spectron's operations.

A new recycling contract was offered to the company in 1980. Because further expansion at Providence now seemed unlikely, the owner of the company bought a second disused paper mill site (zoned for industrial use) at Childs, about three miles down river and within a few hundred yards of several expensive houses. The owners of these determined to oppose the proposed operations. One resident set up a group called Residents for Unpolluted Neighbourhoods which waged an emotional campaign to prevent developments at all costs. Another pressure group, the Little Elk Creek Civic Association, was formed to oppose the facility using more rational argument. This took the fall-back position that, if the plant was installed, it should be subject to stringent environmental controls.

The Air Management Administration of the Maryland Department of Health and Mental Hygiene was subjected to considerable public and political pressure to refuse Spectron a permit to construct at Childs. The principal arguments were the inappropriateness of the site, the likelihood of odours and the previous poor record of the company. While an official admitted that the site was 'still not a good location' and acknowledged Spectron's poor reputation, the agency insisted that location was a land use matter for the county to decide. Since total emissions under normal conditions were only expected to be around 0.5kg per hour, the permit was issued subject to various conditions, some (including a requirement to install an activated carbon filter to reduce odours) inserted as a result of opposition from the tandem residents' groups.

Spectron felt that, since the Childs site was zoned for industry, no county land use permit was required. However, the county Office of Planning and Community Development insisted that a site plan containing a large quantity of information be submitted before a zoning certificate could be issued. Spectron appealed to the county Board of Appeals against this decision and was opposed by both the county and the residents, who also acted as liaison between the state and county agencies and ensured adherence to the relevant regulations. After a heated hearing, the appeal was rejected and the residents' association attempted to buy the Childs site. Spectron lodged an appeal to the courts.

It had become apparent, however, that an accommodation would have to be reached with the county if the company was to fulfil its new contract. Accordingly, after protracted and frequently acrimonious negotiations, the owner signed a legal agreement with the county in 1982 to withdraw his appeal, vacate the Providence site by 1993, forego the Childs site and build a new one on an industrial park – in return for $3M in industrial revenue bonds which provided Spectron with low cost financing. The long-running pollution saga, which appeared to be resolved to the satisfaction of most of those involved, was however not yet over as the developer petitioned the county in 1984 to release Spectron from the closure date stated in the agreement. Needless to say, the county refused. Complaints have continued to be received and the state air management administration has had to take several further enforcement actions. Regrettably, the evangelically zealous chairman of Residents for Unpolluted Neighbourhoods could not accept that the pressure groups' legitimate opposition was bearing fruit. He was indicted on a charge of paying an undercover police officer to destroy the buildings on the Childs site just before Spectron's agreement with the county was signed.

Notwithstanding the earlier closure of the Providence site on air pollution grounds, the different degrees of discretion available under air pollution control and land use planning legislation can be seen clearly in this case. The air construction permit for the Childs site was amended to include extra conditions but the land use planning ordinance was interpreted to mean that the application should be refused on the largely procedural ground that no site plan was submitted. It was this refusal, and the use of similar discretionary powers at the Providence site, that paved the way for the negotiations leading to the agreement to relocate. While the burden of proving that pollution from Spectron's plant was causing cancer was beyond the state, the county was more prepared to accept that the relationship existed, or at least that the local residents believed it existed. The county arranged a relocation of the plant partly, no doubt, to retain the employment and property tax revenue benefits it conferred but largely to accommodate its persistent owner. The new location should, at the very least, allow much better pollutant dispersion than the confined valley site at Providence.

THE OREGON ENERGY RECOVERY FACILITY

The Portland Metropolitan Service District (Metro) obtained land use and air pollution permits to construct a resource recovery plant in Oregon City, near Portland, in 1977 and 1980 respectively, without difficulty. The EPA PSD permit specified that particulate emissions should not be more than 0.04 gr/ft^3. Following a reorganisation of Metro, the design of the facility changed to energy recovery, an agreement being signed with the Publishers Paper Company to deliver steam via a pipeline. The 4 ha site lay next to a landfill and about a quarter of a mile from numerous residential properties. The plant was to be privately constructed, owned and operated; Wheelabrator-Frye Inc were chosen as the operating company.

Metro applied to Oregon City for a conditional use permit[8] for the energy recovery facility in October 1980. A joint meeting of the Oregon City Planning Commission and the Zoning Board of Adjustments was held in November, at which several members of the public spoke. More information was demanded. An independent report was commissioned which concluded that there would be no significant air pollution impacts and that 'present data on dioxin emissions is not sufficient to curtail the proposed energy recovery project'. The planning staff of Oregon City prepared their own report and recommended conditional acceptance of the project. Following three further joint meetings, at which vociferous opposition to the plant was expressed and petitions presented, the appointed planning commissioners voted unanimously to grant permission subject to 25 conditions.

The objectors appealed to the elected city commission against this permission. Rival opponent (Oregonians for Clean Air) and proponent (Citizens for Common Sense) groups were formed and two further lengthy meetings ensued at which evidence for and against the proposal was presented by numerous speakers. Eventually, the permit was approved 4–1, the lone vote of protest being cast by a commissioner who had become a dedicated opponent of the scheme, partly because of his fears about the effects of dioxin emissions. Some 29 conditions were appended, including nine relating to air pollution control. One of these, conceded by Metro, required a scrubber as well as an electrostatic precipitator to be fitted to reduce emissions of sulphur dioxide and other acidic gases by about 75%. Another related to payments in lieu of property taxes if Metro, rather than a private company, operated the facility. The objectors now appealed to the Oregon Land Use Board of Appeals against the unenforceable nature of certain of the permit conditions and against procedural improprieties. However, because they had not lodged their appeal within the requisite 30 days, it was dismissed in November 1981.

The application for a permit to discharge air contaminants was lodged in June 1981. It ran to several hundred pages and proposed that emissions of particulates (at not more than 0.02 gr/ft^3) would be about 150 tons per annum and of sulphur dioxide about 600 tons per annum. The treatment of the application by the Air Quality Division of the Oregon Department of Environmental Quality was complicated because, whilst the whole area was in attainment for sulphur dioxide, parts were not in attainment for particulates, ozone and carbon monoxide. Sulphur dioxide would thus be subject to PSD regulations administered by EPA but particulates and hydrocarbons would be subject to the more stringent non-attainment regulations and offsets would be required. The air quality division demanded more information and re-modelling.

Meanwhile, Oregonians for Clean Air organised a ballot in Oregon City against the facility but lost narrowly. Changes to the project took place and, to avoid being subject to power plant siting regulations (more than 25 MW was now to be generated), Publishers, the electricity user, obtained a special dispensation from the state senate. EPA reported that dioxin from incinerators posed no human health risk and its PSD permit was granted in June 1982. The air quality division determined that the lowest achievable emission rate for particulates was 0.015 gr/ft^3, giving 84 tons per annum. Metro agreed to provide offsets of 10 tons by landfill closure and 74 tons by operating a burnable garden rubbish collection scheme at an annual cost of $70,000, adjusted for inflation. Redefinition had shown that the locality was not, after all, subject to hydrocarbon offsets. The state weakened some of its other proposed conditions at the developer's request.

A long established pressure group, the Oregon Environmental Council, was now pressing for a fuller analysis of the effect of sulphur dioxide emissions on state standards. It was attempting to raise control costs to the point where the developers would be forced to relocate their plant. A lengthy public hearing into the proposed state permit was held in July 1982 at which the problems of securing offsets, sulphur dioxide control and the health effects of dioxin were three of the main issues. The state consequently demanded 80% scrubbing efficiency, reducing sulphur dioxide emissions to 168 tons per year in its revised draft permit. The developers claimed the plant would have the highest level of pollution control of any energy recovery facility in the world. Several design changes were thus made after the grant of the land use permit.

The one dissenting city commissioner had meanwhile obtained enough names to have the energy recovery facility placed before the electorate in Oregon City and in the surrounding area once again. The neighbouring City of West Linn formally voted against the project. Publicity campaigns involving canvassing, press, radio and television were mounted by both sides, with impassioned and not always accurate statements being made by Oregonians for Clean Air. The Oregonian newspaper had been printing articles sympathetic to the objectors' position for some time. This time the vote went against the facility, even in Oregon City. Metro felt it had no alternative but to withdraw the scheme and revert to landfill elsewhere despite an expenditure by Wheelabrator-Frye and itself of over $2M on the permit application process. The city commissioner had failed to prevent the grant of the land use permit and the grant of the air contaminant permit was in train, but this one man succeeded in killing the project on an initiative ballot, despite the substantial reductions in emissions conceded during negotiations.

This case demonstrates the complexity of the air pollution permit process, with the state both employing models and endeavouring to maximise the reductions of low level emissions in its particulate offset arrangements. It also illustrates the discretion available to air pollution controllers, as conditions were adjusted at the behest first of the developer, and then of the objectors. In the end, the air pollution control conditions would have been much more stringent than those imposed in 1980, when little controversy had arisen. The financial attractions of the energy recovery facility had persuaded most Oregon City representatives to approve the project and they ensured that revenue in lieu of taxes would be received if Metro operated the plant. While the city was thus bound to seek mitigation of impacts and not refusal in its land use permit process, its involvement in setting air pollution control conditions appears to have sprung directly from the nature of Oregon's comprehensive land use planning system. The uncertainties and fears associated with dioxin pollution undoubtedly helped the objectors to persuade the local population to vote against the project.

THE CALIFORNIA REFINERY MODIFICATION

Chevron Inc has operated a refinery in Richmond since 1902. Chevron proposed to modify its existing plant, which is a very major pollution source in the San Francisco Bay area, in order to upgrade its output and add flexibility of operation. Chevron believed it needed permits only from the City of Richmond[9] and from the Bay Area Air Quality Management District. At first Richmond did not feel that an environmental impact report under the provisions of the California Environmental Quality Act would be required because no discretionary permit appeared to be involved for a use classified as non-conforming in the city zoning ordinance. Chevron, however, was anticipating opposition to its project and was determined that no procedural objection could be sustained in the courts, and it was agreed to require an environmental impact report before a minor traffic permit could be issued.

Chevron applied for this traffic permit in September 1980 and submitted a project information form putting the cost at $390M with a construction workforce of 1,200 and a permanent workforce of 25 (later raised to 45). It was apparent that the main topic in the environmental impact report would be the effect of air pollution.

An agreement was reached with EPA and an application for authority to construct a pollution source was submitted to the air quality management district in March 1981, following preliminary discussions. The 200 page document was eventually deemed complete, following the provision of more information and $11,000 in fees, in August 1981. Chevron's position was that, because of emission reductions already banked or anticipated from the modification, there would be no emission increase and hence no need for complex modelling, and that no additional throughput would be involved. The district felt that a 25% increase in throughput, and hence emissions, was possible and that conditions would be necessary to control these. Eventually, the two sides agreed that a bubble or 'cap' covering most of the refinery operations would be utilised and that no net increase in annual emissions would ensue. A formal 'summary of analysis' was released in November which drew criticism from Citizens for a Better Environment, the local branch of a national pressure group, on the grounds that higher hourly, daily or weekly emissions might result and that the permit resulted from negotiation, which was inappropriate.

The district decided to modify its objective to that of a net decrease over the year by demanding offsets in a ratio of 2:1 against the annual base line emissions for refinery and wharf operations. A further summary of analysis was issued in March 1982 specifying emission reductions which, though only a very small percentage of overall emissions, were mostly considerably greater than the figures originally advanced by Chevron. Despite further protests from Citizens for a Better Environment about the lack of hourly, daily and weekly emissions limits, the authority to construct was issued in April 1982.

Scoping for the environmental impact report started in February 1981 and the draft, prepared by consultants, was published in December 1981 and distributed for comment. A public hearing was held and the final report (which cost about $200,000) was released in March 1982. Citizens for a Better Environment raised numerous objections on the air quality section both in writing and at a public hearing before the City of Richmond Planning Commission. It had now been determined that a conditional use permit under the zoning code would be necessary and this was granted by the commission after Chevron had agreed to contribute $750,000 for traffic management measures and road plans. Conditions were also appended relating to road sprinkling and other matters raised in the environmental impact report, but not to air pollution emissions. Citizens appealed against the grant of this permit and, at a City Council hearing, stated that the project was a 'bad deal for air quality' and that the opportunity to control total emissions from Chevron had been squandered. However, once again, Citizens lost and the appeal was denied by the council at the end of March 1982. Citizens decided not to appeal to the courts as it felt it had a better chance before the Bay Area Air Quality Management District Hearing Board.

Despite several court cases originated by Chevron's team of lawyers to deny the pressure group access to the board, Citizens appealed against the district's permit decision and the hearing commenced in May 1982. Only the pressure group's argument about increases in hourly, daily and weekly emissions was admitted by the board, which rejected the appeal in March 1983. The board felt that the district's rules allowed sufficient discretion for the permit to be procedurally correct. Citizens' application for a rehearing was refused and no appeal was made to the courts as the group's manpower was by now over-stretched. The modification project was completed in 1985.

Chevron stated that the new approach adopted by the district to bubbles penalised the company but most observers felt that, partly because this was a relatively uncontroversial modification project, Chevron had been awarded a generous permit, even if it had conceded further emission reductions during negotiations. Without the involvement of Citizens for a Better Environment, however, it seems likely that only the original, no net increase, permit would have been obtained by the district and Chevron would not have made its later concessions. There was clearly enough discretion available to have obtained more stringent conditions. There was never, at any stage, the slightest hint that either the City of Richmond permits or the Bay Area Air Quality Management District permits would not be forthcoming. The attitude of the city to this important source of employment was indicated by its omission to recognise the need for a use permit. Negotiations (which appear to have been quite proper, despite Citizens' protest) were always geared to mitigation of impacts. These led, among other concessions, to the contribution to road plans and traffic management measures. The whole permitting process must have cost Chevron well over $1M. Though the outcome was never in doubt, the opportunities for protagonists to appeal and to take legal action were still very evident.

THE CALIFORNIA CHEMICAL PRODUCTION FACILITY

In early 1975 Dow Chemical Company unveiled plans to build a $500 million chemical production facility on the banks of the Sacramento River, 35 miles north east of San Francisco[10]. Much of the complex was to be built on a 1,100 ha site in rural Solano County, and connected by submerged pipelines to the rest of the complex to be constructed in the industrialised town of Pittsburg, just south across the river in Contra Costa County where Dow had operated another chemical plant since 1940. The new works, designed to produce basic chemicals such as styrene, vinyl chloride and ethylene was expected to employ 1,000 construction workers and 800 permanent staff. The site was classified as non-attainment for particulates, hydrocarbons and ozone.

Solano County, a small, fast-growing jurisdiction calculated that Dow would add 14% to its assessed valuation without bringing in many new residents. It was therefore much in favour of the project. Dow needed a total of 65 permits and approvals, including rezoning of the prospective site from agricultural to industrial use; cancellation of an open-space designation and certification of an environmental impact report by the county and numerous air emission permits from the Bay Area Air Pollution Control District (now called the Air Quality Management District).

Dow engaged local consultants and lawyers. The consultants started work on the environmental impact report in 1974 and, in December 1975, the county certified the report, despite reservations expressed by several state agencies. The county zoning board then rezoned part of the site from agricultural to industrial use and cancelled a contract preserving the rezoned land for agricultural use. Next, the county started drawing up a specific industrial development plan for the area. However, Friends of the Earth, the Sierra Club and a San Francisco group called People for Open Space appealed against this decision to the county court, fearing that the rezoning and other county actions, which they believed to be contrary to California law, would establish damaging precedents.

On 4 May 1976, after months of negotiations, Dow formally submitted its request to the air pollution control district for permits to build the styrene plant. The district, though it had refused some permits in the past, had a reputation for flexibility in negotiating permits within the compass of its rules. The main pollutants were particulates, sulphur dioxide, oxides of nitrogen and hydrocarbons. Dow's air pollution control equipment was deemed to be more than adequate but, following modelling, it was determined that emissions of particulates and hydrocarbons exceeded the significance thresholds set in the district's regulations current at the time and the district, therefore, had to deny the permits under a rule which prohibited authorisation of any new facility 'which may cause the emission or creation of a significant quantity of any air contaminant which would interfere with the attainment or maintenance of any air quality standard' (Duerkson, 1982:54). A preliminary denial was issued on 8 July 1976, and a public hearing into Dow's appeal took place on 19 July in Solano County.

Over 600 people attended this noisy hearing, most of them in favour of the project. However, it was explained that the district had no leeway to grant the permit under its rules. One moderate business leader in the Bay Area argued that 'economic and political factors deserved equal weight with air quality. The air district in effect has become the land-use agency for the Bay Area and probably put the lid on economic development' (Duerkson, 1982:64). Notwithstanding, the district reaffirmed its earlier denial on 12 August. Dow had been a victim of the basic difficulty of articulating a balance between air pollution control and development inherent in the Clean Air Act.

Dow appealed against this decision to the district hearing board. Just before this was heard, the district's directors reconsidered the rule causing the difficulty but refused to amend it pending further study. The appeal hearings went on for several weeks and were technical and legalistic. They were adjourned on 28 October 1976, additional hearings being scheduled for the near future.

Dow also had difficulties obtaining the necessary approvals from state agencies which, even though they had already responded to the environmental impact report, continued to ask for further information. A federal environmental impact statement was generating further questions. Governor Brown called for a consolidated hearing on Dow's applications for state permits and directed state agencies to submit a final list of questions about the project. These meetings resulted in the prospect of having completely to rewrite the environmental impact report and the company cancelled the project on 18 January 1977, blaming the state for bureaucratic delays and environmentalists for obstructionist tactics. The court case, the air permit hearings and the federal environmental impact statement were never completed. It is not clear, however, that Dow did not have its own additional reasons for cancelling the project.

In California, the effect of this decision was immediate and the regulatory pendulum swung back towards development. At the national level, the Environmental Protection Agency adopted the offset policy to accommodate industry in dirty air areas. The Bay Area Air Quality Management District adopted similar regulations and, were Dow to have reapplied, it could have reduced emissions at its existing Pittsburgh plant to provide the offsets for the air pollution from its new complex and thus been granted its permit.

This case remains a very unusual instance of a serious developer being refused an air pollution permit. It is also interesting because of the attempts by various state agencies to investigate the impacts of the development further, after they had agreed to the relevant sections in the environmental impact report, which obviously did not serve its desired anticipatory purpose. While most state officials were seeking mitigation of environmental effects, some were undoubtedly seeking delay or refusal. Solano County, because of the revenue and employment benefits, had precipitately granted the necessary local land use permits. The involvement of several major groups of objectors in a wide variety of permit procedures and appeals, which afforded adequate opportunities for frustrating Dow, was notable.

NOTES

1 See, for example, Hagevik et al (1974: Chapter 2), in which it is claimed that the provisions for preconstruction review in the Clean Air Act are essentially concerned with land use regulation.
2 Similar arguments have been advanced about new source performance standards: that the discrimination against the new may be a sophisticated form of protectionism by certain legislators and business interests. See Ackerman and Hassler (1981) in which the success of the coalition between eastern business, miners and environmentalists to impose uniform scrubbing standards on all power plants (whether burning low or high sulphur fuel) is criticised.
3 It is, of course, possible and indeed likely that some developers never reach the formal application stage when they learn the nature of the requirements demanded of them.
4 *Clean Air Act 1970*: 110(A) (2)(B).
5 A longer account of this case study is to be published (Wood, forthcoming).
6 For a well-researched account of the passing and implementation of this act, see DeGrove (1984). See also Healy and Rosenberg (1979).
7 For accounts of the legislative response to the problems of burgeoning development in Florida see Healy and Rosenberg (1979), DeGrove (1984) and DeGrove and Stroud (1987).
8 Oregon has enacted particularly strong land use controls. See Leonard (1983) and DeGrove (1984).
9 For accounts of the planning and protection of California's coast see Healy (1978), Healy and Rosenberg (1979), Mazmanian and Sabatier (1983) and DeGrove (1984).
10 This account is derived from Duerkson (1982). See also Storper et al (1981) and Blowers (1984).

5
Preventive controls over air pollution in the United Kingdom

This chapter, which follows much the same pattern as Chapter 3, is devoted to an explanation of the systems of air pollution control and land use planning in the United Kingdom. The first section analyses the context for environmental regulation generally. In the next part of the chapter, air pollution trends are briefly described and the UK legal control framework is outlined. As in Chapter 3, the legal framework of land use control is reviewed and the provisions relating to development plans and development control are explained. The next section includes a discussion on the extent to which air pollution goals have been considered in planning practice. Finally, there is a section on environmental impact assessment.

CONTEXT

Physical characteristics

The United Kingdom is only about the size of Oregon (244,000 square kilometres) but had a 1986 population of over 56,500,000, almost a quarter of the USA's (Central Statistical Office, 1988:6) giving a population density nearly 10 times that of America (229 persons per square kilometre). Thus, in comparison with the USA, Britain is a densely populated little island.

The *stewardship ethic* which has long prevailed in the United Kingdom is in marked contrast to the US frontier ethic. The scarcity of land and the cultural appreciation of the British – particularly the English – landscape and historical heritage has led many property-owners to acknowledge a *public trust* component to their ownership of land. The aristocratic concept of managing land to leave it enhanced, or at least not degraded, for the enjoyment of heirs is quite different from the American land ethic. The result is that many property owners have been willing to accept stringent restrictions on the use of their land in the public interest. The postwar planning legislation of the late 1940s, which was genuinely radical, thus enjoyed widespread support from most sections of the population.

In the United Kingdom, about 19% of land is in public (central and local government plus statutory undertakers) ownership (Massey and Catalano,

1978) but a public right of access exists only over a small proportion of this land. Although historic footpath rights exist in Britain, the desire to obtain a greater degree of public access to areas of outstanding scenic beauty while preserving private agricultural and estate management interests was one of the reasons for the setting up of the British national parks and the growth of land use controls. Another reason is the fact that very little of the UK land is neither cultivated (even if this means grazing or forestry) nor in urban use. Prevention of change of use from agriculture in a country with experience of food shortages in two world wars became, perhaps understandably, almost a national obsession.

The feudal history of Britain, which eventually led to the stewardship ethic, has also made British society more hierarchical and class conscious than American society. It is also far more culturally homogeneous, despite its recent racial problems: persons of West Indian, Indian, Pakistani and African birth only constitute about 3% of the total population[1]. In short, as Corden (1977) put it:

> *Relative to the United States*, the United Kingdom is a society which appears to exhibit a high degree of consensus and agreement on certain basic values as well as mechanisms for engendering acquiescence (p 19, emphasis in original).

In contradistinction to the United States, the homogeneity of British society has led to a concern with product or ends rather than process or means. This may be seen in the concept of fixed target populations in the master plans for cities, in the emphasis in planning on the physical distinction between town and country and in the discretion allowed to appointed or elected representatives to take ad hoc decisions on land use and air pollution control.

It has also led to a *unitary* view of the public interest. As Hall et al (1973) stated, 'this is opposed to a utilitarian view, which would essentially consist of trying to compare and sum up individual pleasures and pains' (p 69). The unitary or *organismic* view of the public interest implies the existence of central decision makers who are considered particularly qualified to adjudge common ends and according to Meyerson and Banfield (1955):

> who can perform the largely technical function of adapting means most efficiently for the attainment of these ends, and who have the power to assert the unitary interest of the 'whole' over any competing lesser interests (p 327).

This view helps to explain both the greater acceptance of land use controls and the greater propensity to design environmental protection regulations which leave considerably greater discretion to the individual official in the UK as compared with the US.

Climate

The temperate climate of the United Kingdom is less extreme than that in most parts of the United States. Britain lies north of the most northerly point of the 48 conterminous states; it is thus cooler and evaporation and transpiration are lower. An annual rainfall of less than 100 cms, which is typical of most of Britain, is considerably less than that in much of the eastern United States. This rain falls more gently than in America and, since vegetation thrives, soil erosion is much less serious (Clawson and Hall, 1973:230). Temperature ranges are also much smaller and the seasons are far less clearly demarcated than in the US.

The climate of Britain partly explains the British attitude to air pollution. London, which stands in a shallow basin, became notorious in the age of Dickens for pea-souper fogs caused by smoke being trapped beneath inversion layers. However, in comparison with the United States, Britain is a small windswept land, surrounded by water, where inversions are much less common. Summer temperatures are rarely high enough to cause serious ozone problems and the wind-assisted export of pollutants emitted through high chimneys has for long been an accepted means of control. Recently, however, the complaints of Scandinavia and (of more local political significance) Scotland over acid rain have caused this policy to be disputed and modified (Weidner, 1987).

The urban tradition in the UK, which has been rather different from that in the US, provides another reason for differing perceptions. The proximity of industry and housing in densely populated cities has led to an acceptance of industry by the residents. The UK air pollution problem has historically been mainly domestic, rather than industrial, in origin and has been recognised as such by the majority of the population. While industrial pollution problems exist, of course, there is no general belief that industry is the sole cause of pollution. Nor is there widespread suspicion of business. Many people remember how bad smog (smoke plus fog) used to be and remember changing from coal fires to cleaner fuels. They participated in the improvement they witnessed.

These factors lead to a crucial difference in view about air pollution and its control between the USA and the UK. The British appear to accept that the consumption of goods, as well as their production, causes air and other types of pollution (see Figure 3): the Americans seem very reluctant to recognise this.

Government system

The United Kingdom of Great Britain and Northern Ireland is a constitutional monarchy consisting of four countries: England, Wales, Scotland and Northern Ireland. While there exist, for executive purposes, the separate ministries of the Department of the Environment (created in 1971 to centralise control over environmental matters in England), the Welsh Office, the Scottish Development Department, and the Department of the Environment for Northern Ireland, each responsible for a wide range of functions relating to the physical environment (including much pollution control and land use planning), the policies these ministries execute are made by the central government in legislation approved by Parliament. Central government retains much greater control than in the US, there being no subsidiary legislative authorities equivalent to the states. A significant consequence of this centralisation is that much air pollution control is undertaken directly by a central government body, Her Majesty's Inspectorate of Pollution.

Local government in the United Kingdom is the responsibility of elected local authorities which provide services under specific duties or powers laid down by Parliament and subject to a high degree of central government control. In England there are 364 district authorities, with an average population of over 100,000, and 39 county authorities.

Outside the metropolitan areas, every square metre of land falls within the administrative areas of both a district and a county. Broadly speaking, the counties are responsible for strategic planning and such matters as refuse disposal whereas the districts are responsible for the vast majority of planning control and some types of air pollution control. The local government arrangements vary in Wales, Scotland and Northern Ireland (Byrne, 1986).

Local government now receives over half its income as grants from central government. This not only increases the dependence of local government upon the centre but means that much of the benefit it may receive in increased rates or other local property taxes from a new source of industrial air pollution will frequently be lost by an adjustment of its grant from central government. For this reason, and because local authorities in Britain are large, the direct revenue benefits of constructing new industrial premises to local communities are not very great. Job creation is a central objective of most local authorities, however.

Contrary to the general view of the integrity of the British civil service, local government is not without taint of corruption, though the scale of such problems has certainly never been remotely comparable to American city graft. Generally, however, the suspicion of government so prevalent in the USA has not extended to Britain where local governments are considerably larger and less parochial.

In both the US and the UK there is an intimate relationship between public and private power or economic and political management. This is perhaps strongest in the UK but in both countries it can act, as Enloe (1975) stated, as a counter 'against taking decisive action to limit man-made environmental hazards' (p 318).

Environmental regulation

The United Kingdom, in marked distinction to the United States, uses broad enabling legislation and flexible rules based upon various local environmental, economic and technological factors. This system, which is not open to such significant public scrutiny as in America, permits much greater freedom to choose control methods but requires a degree of mutual trust and co-operation between the regulatory officials and regulated firms that may not always be present. As Peacock (1984) observed:

> Legislation is usually couched in language which implies a fairly wide scope for discretion and interpretation on the part of the agents empowered to make and enforce regulations. The words 'as far as is reasonably practicable', which appear with such regularity, are obviously intended to provide a degree of latitude to both the regulators and the regulated (p 96).

The main permit the developer of a new air pollution source needs to acquire is planning permission from the local authority and this is the only process which is open to significant public scrutiny at present. Negotiations between the developer and pollution control authorities over the very limited number of permits required normally take place in private. While planning appeals are reasonably common there have been very few environmental court cases, and transaction costs for developers are usually small.

Environmental regulation has been relatively uncontroversial in Britain. There have been no environmental regulation control debates of the type that have raged in the United States, though the recent Conservative government has endeavoured, by advice and by the dismemberment of advisory bodies, to swing the pendulum towards development and away from environmental protection. Industry and the general public appear to be largely satisfied with the current system, though certain interest groups are not.

With the exception of some of the reports of the Royal Commission on Environmental Pollution, there has been a notable absence of influential proposals to improve the efficiency of British environmental policy. One reason for the degree of public satisfaction with environmental control is that the land use planning legislation frequently provides opportunities both to debate and to control pollution. Changes have, however, recently been instigated or proposed in both the air pollution and the land use control systems.

AIR POLLUTION CONTROL

Trends

The United Kingdom, like the United States, has experienced a significant improvement in air quality in recent years: at least so far as smoke – sus-

96 *Planning Pollution Prevention*

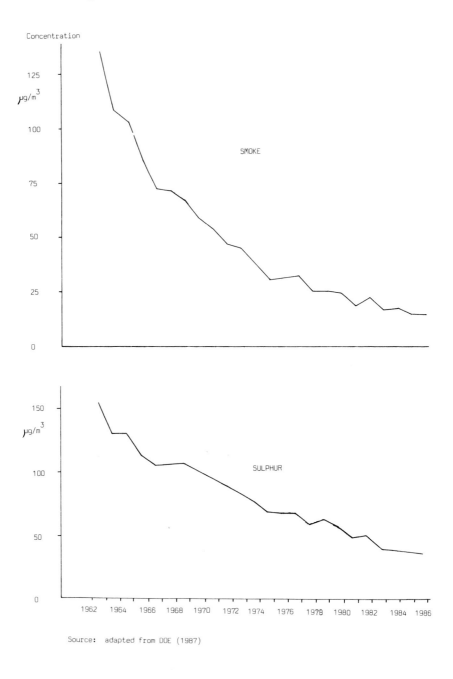

Figure 6 UK air quality trends 1962–1986

pended particulates of less than 15 micrometre diameter – and sulphur dioxide are concerned. Figure 6 shows the trends in the concentrations of these two pollutants. While there has been some dispute as to the precise contribution of the smoke control area provisions of the 1956 and 1968 Clean Air Acts, there is no doubt that changes in domestic heating methods have been instrumental in bringing about the reductions in concentrations observed. There are now more than 6,000 smoke control orders covering almost two-thirds of urban properties (DOE, 1987b). The average concentration of smoke in 1986 was less than half the 1977 level and about one-tenth that prevailing in 1962. Similarly, sulphur dioxide levels in 1986 were about 60% of 1977 concentrations and about 25% of those in 1962, reflecting a reduction in emissions from low level sources (DOE, 1987b).

Another influence on concentrations has been the replacement of obsolete housing stock in the inner cities by modern residential development with smokeless forms of space heating which leads to de facto smokeless zones. Large-scale urban renewal has also afforded an opportunity to remove long-standing bad neighbours such as animal treatment works, dye works, tanneries etc, from residential areas.

Industry has also made its contribution to amelioration. There remain, of course, isolated pockets of high concentrations of smoke and sulphur dioxide, but attention in recent years has begun to focus on the health effects of specific pollutants (notably lead), on acid rain (Weidner, 1987) and on other previously unmonitored pollutants from industrial and vehicular sources. Elevated concentrations of ozone had been detected on hot days in London and other parts of Britain, indicating that photochemical smog can no longer be ignored (Photochemical Oxidants Review Group, 1987). However, the levels have generally been lower than those recorded in the United States and, as yet, the only controls over mobile sources of air pollution have related to the lead content of gasoline and the opacity of exhaust fumes.

Legal framework

Like the United States, the United Kingdom has suffered from a number of notorious air pollution incidents. Early problems from the burning of coal in London led to several prohibitions and other attempts to control smoke. Further, rather ineffectual, measures were enacted in the 19th century, aimed at reducing the smoke problems experienced in numerous industrial cities. The Public Health Act dates from 1875, though more limited powers had been enacted earlier. Emissions of hydrogen chloride from alkali works on Merseyside led to damage to grasslands and cattle in the middle of the 19th century. This could not be controlled under the existing inadequate smoke control legislation and the first Alkali Act was passed in 1863 to cope with the problem.

TABLE 6 UK national ambient air quality standards

Pollutant	Limit value
Smoke	Annual (median of 24 hour values) 80 μ/m^3
Sulphur dioxide	Annual (median of 24 hour values) 120 μg/m^3 if smoke < 40 μg/m^3 80 μg/m^3 if smoke > 40 μg/m^3
Nitrogen dioxide	Annual (98th percentile of hourly concentrations) 200 µg/m^3
Lead	Annual (arithmetic mean) 2 μg/m^3

Source: Haigh (1987)

Nearly a century later, the celebrated London smog of 1952 was estimated to have led to some 4000 excess deaths due to the exacerbation of bronchitis, emphysema and other lung diseases. The Beaver committee on air pollution was set up as a consequence of this catastrophe and the Clean Air Act 1956 was the direct outcome of its deliberations[2]. The Control of Pollution Act 1974 was, at least as far as air pollution control was concerned, largely a codifying measure. As in the United States, anticipatory control over stationary sources is fragmented among various legislative provisions relating to pollution. There is some, partial, control over pollution from motor vehicles and Commission of the European Communities air quality standards for suspended particulates, sulphur dioxide, lead and nitrogen dioxide are in force (Haigh, 1987). These standards are shown in Table 6. It is notable that they are considerably less stringent than their American counterparts (Table 4) but, as in the USA, these standards are not infrequently exceeded. (The deadlines for compliance vary between 1989 and 1994.)

Created by an Act of Parliament in 1863, the body now styled Her Majesty's Inspectorate of Pollution (HMIP), previously the Industrial Air Pollution Inspectorate, Her Majesty's Alkali and Clean Air Inspectorate and before that the Alkali Inspectorate, has come under the jurisdiction of various central government departments. Since 1987 it has been assigned to the Department of the Environment, having been one constituent of the Health and Safety Executive (HSE). Unlike its predecessors, which were concerned only with air pollution control, HMIP is also responsible for the control of some water pollution and wastes and for determining the best practicable environmental option (Chapter 1) for certain industrial processes.

Registered stationary sources
A principal function of HMIP is the enforcement of the consolidating Alkali, etc, Works Regulation Act 1906, as amended by successive laws, orders and regulations[3]. The purpose of this recondite body of legislation has been the control of certain noxious atmospheric emissions from a number of specified chemical, metallurgical and other industrial processes. The processes which have been added to the Inspectorate's jurisdiction are those which, by virtue of their complexity in terms of air pollution control, are considered to require a greater degree of technical expertise than can generally be expected to be possessed by local authority environmental health departments (the bodies responsible for the control of non-registered sources – see later). Thus the sources emitting the largest quantities of, and the most active, pollutants are registered with HMIP: power stations, oil refineries, most chemical works, fertiliser plants, most lead works, many smelters, etc.

An act of 1874 introduced what is now the crucial concept of *best practicable means* (BPM); namely, measures which operators of alkali works had to take to prevent or limit the discharge of certain specified noxious gases. For all but two of the 60 or so processes listed in a much amended schedule to the 1906 Act[4], no statutory limit on emissions is involved and control is effected solely by requiring the operators of such scheduled processes to use:

> The best practicable means for preventing the escape of noxious or offensive gases by the exit flue of any apparatus used in the work, and for preventing the discharge, whether directly or indirectly, of such gases into the atmosphere, and for rendering such gases where discharged harmless and inoffensive.

Neither the Alkali Act, nor any subsidiary act or instrument, gives a definition of the term best practicable means. However, the 1956 Clean Air Act does offer an elaboration of the closely allied term *reasonably practicable*: 'reasonably practicable having regard, amongst other things, to local conditions and circumstances, to the financial implications and to the current state of technical knowledge'[5]. It must be assumed that this interpretation is accepted by the Inspectorate for the Chief Inspector has stated that 'it would be unreasonable to interpret the Alkali Act differently' (DOE, 1974:11).

Historically, best practicable means, as enforced by Her Majesty's Inspectorate of Pollution, has amounted to a pragmatic system of pollution control with account taken of three broad sets of factors: local, economic and technological. The economic and technological factors are brought to bear in setting *industry-wide* BPM. The local factors only come into play in adjusting the industry-wide controls to particular premises, ie, in setting *specific* BPM.

Best practicable means is interpreted sequentially, *prevention* requiring the use of technical controls such as electrostatic precipitators, bag filters, absorbers, etc, and *discharge* requiring the use of chimneys of sufficient height to render the resulting ground level concentrations acceptable. However, reliance on dispersion alone means that prevention is impracticable (HSE,

1984:21). The legislative control requirement is taken to include not only stack emissions but the abatement of fugitive emissions, the regulation of the industrial process itself and the maintenance and proper use of equipment. As well as involving this comprehensive range of source controls, which include *housekeeping arrangements*, best practicable means also includes emission standards, in the form of presumptive limits[6]. These non-statutory standards are set by the Chief Inspector and failure to operate within them may be presumed to be evidence that BPM are not being used (HSE, 1984:14). However, 'if a presumptive limit is exceeded, this does not imply automatically that the best practicable means are not being used' (HSE, 1984:15).

In divising industry-wide best practicable means, the Inspectorate has tended to consider the economic consequences of the costs of controlling pollution as paramount:

> The expression best practicable means takes into account economics in all its financial implications, and we interpret this not just in the narrow sense of the works dipping into its own pockets, but including the wider effect on the community (Ministry of Housing and Local Government, 1967:3).

There has been a shift of emphasis recently, to acknowledge that the benefits of control also have economic consequences (HSE, 1982:15).

The technological factors involved in best practicable means result in changes in controls over time. This leads to the much-vaunted advantage of flexibility of control by gradually increasing abatement as knowledge progresses. Best practicable means has been described as ' ... an elastic band ever tightening as chemical science advanced ... ' (HSE, 1982:16). Full best practicable means is reached when there is 'little or no impact on the community and with no scope for further improvements' (Ministry of Housing and Local Government, 1967:3) but it is often necessary to accept a lower standard. Again, there is now an increasingly overt acceptance that knowledge about the effects of pollution is a relevant consideration.

Once a particular process is, by order of the Secretary of State for the Environment and following a public inquiry, added to the schedule the Chief Inspector, after discussions with representatives of the industry concerned prepares a set of guidelines on the requirements of the best practicable means to be observed in the operation of that process[7]. Recently, other parts of the Department of the Environment, the Trades Union Congress and the Institution of Environmental Health Officers have been consulted. Almost without exception, published notes of best practicable means have been prefaced by the qualification:

> These notes are not claimed to be comprehensive, but they do provide a basis for negotiation between works' management and the Inspectorate. Flexibility is left to meet special local circumstances by consultation. There are likely to be matters revealed during routine inspections, which will need attention to meet ... the Alkali Act (DOE, 1975:92).

It will be apparent that the determination of best practicable means lies at the discretion of the Chief Inspector. Moreover, it has traditionally been the view of the Inspectorate that the superiority of the best practicable means approach over other systems of pollution control: for example, air quality or emission standards, pollution licences or charges, lies in its discretionary character and its flexibility both in formulation and in implementation. It might be argued that the Inspectorate has gradually assumed a degree of discretion in the exercise of its duties and, in particular, the determination of best practicable means, in excess of that originally intended by Parliament. This assertion, of course, could be tested only in the courts. However, it seems unlikely, in view of the extent of consultation with the relevant industry which precedes the determination of the best practicable means of the process involved, that such a test case will ever arise (McLoughlin and Foster, 1982; Miller and Wood, 1983).

The Inspectorate has now accepted that BPM and air quality standards (AQS) can be complementary despite the essentially case-by-case nature of BPM:

> AQS and BPM are therefore perfectly compatible, indeed complementary, in nature ... where the AQS were breached, or in danger of being breached, it would be necessary for consideration to be given not only to the implications for BPM in the particular local circumstances (in terms of more stringent control) but also to the control of sources of pollution outside the Inspectorate's remit, and to the control of possible additional developments (HSE, 1982:17).

This statement is the strongest Inspectorate public utterance on the importance of local circumstances in determining best practicable means or, indeed, of the necessity of using land use controls to prevent pollution (see Chapter 6). Such recognition of the possibility of making best practicable means locally variable has only recently been reflected in other HMIP pronouncements. It appears, however, that Parliament, in debating the bill which became the Clean Air Act, attached considerable significance to the term *local conditions and circumstances*. The inference was that stricter standards would be appropriate where pollution was likely to be particularly serious (Social Audit, 1974:9). This interpretation has been clarified recently with the addition of two more factors in the consideration of best practicable means: current knowledge of the effects of pollutants; and local conditions and circumstances where these are relevant in requiring particularly stringent control (HMIP, 1988).

In deciding on *specific* best practicable means for a particular works, the district inspector, acting on the Chief Inspector's behalf and following discussions with the works' management, specifies exactly what best practicable means for that works shall entail.

> The Chief Inspector, with the help of his deputies, lays down the broad national policies and provided they keep within these broad lines, inspectors in the field have plenty of flexibility to take into account local circumstances and make suitable decisions (DOE, 1974:13).

The controls of HMIP are both retrospective and prospective. It is of particular relevance to the anticipatory control of air pollution, however, that the Inspectorate's powers of prior approval over the industries it controls extend to 'additional or replacement plant and significant modifications to existing plant' (HSE, 1982:14). There is no minimum size for pollution sources falling within the Inspectorate's control.

In practice, local factors have been of only very minor significance in determining levels of control. Thus, the Chief Inspector has made it clear that the Inspectorate's controls were intended to be uniform:

> BPM for one company are BPM for another operating the same process. Thus, where industry-wide BPM have been stipulated, the same requirements are applied consistently between companies, although local circumstances may require some variation on details (HSE, 1982:15).

While local conditions and circumstances could be interpreted to include existing levels of air pollution and the proximity of housing or similarly vulnerable land uses, all the evidence appears to suggest that, if this is the case, it has been true to only a very minor extent. If anything, it has been interpreted to mean the firm's local circumstances (ie, financial circumstances), though this interpretation may be changing. As a Deputy Chief Inspector put it in 1978:

> We have to take the local circumstances into account; the circumstances could be employment circumstances; they could be planning circumstances; and then of course they could be local pollution circumstances (Miller et al, 1980:65).

When the Chief Inspector spelt out the basic philosophy of implementing best practicable means for a new plant, he made it clear that pollution problems can still arise. 'The consequence is that local amenities suffer. Shutting the plant down rarely provides an answer' (DOE, 1974:13). The inference that damage may arise from pollution is very clear. That such damage is not confined to breakdowns is apparent:

> Whilst the policy of avoiding demonstrable public health hazard has been achieved, BPM cannot guarantee that local environments are fully acceptable in terms of amenity and therefore free from public complaints (HSE, 1982:16).

The adjustment of industry-wide best practicable means to specific works appears, in practice, to involve only variation on details, ie, the adjustment of chimney heights:

It must be remembered, of course, that judgements taking into account other factors, such as local conditions, are also important in the final determination of chimney heights (HSE, 1982:16).

Again, as a Deputy Chief Inspector stated:

The suggestion that the Alkali Inspectorate is not concerned with external quality is misconceived. We are very much concerned and are statutorily required to implement this. In most cases we have to do this by adjustment of chimney heights (Miller et al, 1980:68).

Non-registered stationary sources
While the scheduled processes pose the more technically intractable problems of control, it must be remembered that the 2,000 registered works in England and Wales are greatly outnumbered by the 30,000–50,000 industrial premises not covered by the Alkali Act and from which significant quantities of pollutants are emitted to the atmosphere (RCEP, 1976). There are two such types of operation: those involving combustion processes and those employing non-combustion processes.

Where sources of pollution are not registered but involve *combustion* (ie, industrial boilers and furnaces) the Clean Air Acts grant local authorities some anticipatory powers to forestall pollution from certain non-domestic furnaces. As well as the right to approve the height of any chimney, there is, in some cases, a power of prior approval of equipment to regulate the emission of grit and dust[8]. A local authority must not approve the height of a chimney unless it is satisfied that such height will prevent the smoke, grit, gases or fumes emitted from being prejudicial to health or a nuisance (Webster, 1981). The personnel responsible for administering these controls are environmental health officers, responsible to a committee and thus to the full elected council of the relevant local government district.

Controls over discharges from combustion processes under the Clean Air Acts are not as comprehensive as those over scheduled processes under the Alkali Act. In particular, a local authority has no power specifically to limit emissions of sulphur dioxide and other gases[9].

Where the functions of the Inspectorate overlap with those of a local authority in respect of any premises then overall control is left to the Inspectorate. In these cases the Inspectorate deals with smoke, grit and dust as well as with noxious and offensive gases (Webster, 1981).

Where a process giving rise to atmospheric discharges is neither scheduled nor involves combustion then (save for the right to approve the height of any chimney installed under the Clean Air Acts and to approve the establishment of trades designated as offensive in the Public Health Act, 1936) there is no power of prior approval available under the pollution control legislation for measures to reduce or render harmless any discharge to the atmosphere.

There are numerous such *non-combustion* stationary sources of atmospheric pollution which lie outside the scope of both the Alkali Act and the Clean Air Acts. Certain retrospective powers are available to local authorities under the nuisance provisions of the Public Health Acts once a nuisance or a danger to public health has arisen but they are notoriously difficult to enforce. Sometimes, only closure of, for example, animal treatment works, utilizing these acts can obviate odour nuisance (DOE, 1976). However, this retrospective power is clearly inferior to the right to intervene at an earlier stage and to require the installation of equipment, the observance of operating procedures and the limitation of emissions which ensure that the risk of such nuisance is reduced to a minimum.

Practice and reform

As in the United States, the public in the United Kingdom has consistently seen air pollution as a major problem and supported efforts to control it. Unlike in the United States, however, the control of air pollution has not been a major political issue in recent years. True, HMIP has been attacked as being industry's ally, secretive and remote (Bugler, 1972; Social Audit, 1974) but these accusations have not been supported by the public at large: most people in the United Kingdom have never heard of the Inspectorate. Control by local authorities under the Clean Air Acts and the Public Health Acts is even less controversial.

The general improvement in the levels of the traditional air pollutants has been so marked that complaint by the general public might seem churlish. There is also a widespread lack of knowledge about where complaints should be directed. The split in responsibilities between the Inspectorate, the district council environmental health and planning departments also makes for confusion among the public. Generally, there is a lack of public participation in decisions involving new sources of air pollution, and a marked lack of opportunity for the public to appeal against them.

HMIP is a very small body with, in 1986, 37 air pollution inspectors in post in England and Wales[10]. Each scheduled works was visited, on average, just over five times in 1986 (HSE, 1987). Inspectors are all graduates in chemistry or chemical engineering with at least five years industrial experience. None has any experience in economics, biology or any of the scientific disciplines concerned with the assessment of air pollution damage. The Inspectorate is politically very remote, being answerable to the Secretary of State for the Environment.

Criticisms of the Inspectorate have been persistent, if never politically significant. Rhodes (1981) believed that the degree of discretion accorded it, which is much greater than its American counterparts enjoy, leads it to consider questions which are essentially political in character:

> An inspectorate which interprets its role as being to judge the balance at any given moment between the technical means of controlling pollution and the cost to industry of doing so, which aims to lead industry along the road of gradual progress rather than to state absolute requirements which industry must somehow meet or face the prospect of prosecution ... is extremely vulnerable to misunderstanding and suspicion of its motives (p. 152).

While it is an offence to operate a works in default of the agreed best practicable means, it has become the tradition of the Inspectorate to prosecute only in the event of the most flagrant and persistent abuses. Usually the Inspectorate considers it sufficient to issue an admonition in the form of an *infraction letter* indicating the nature of the violation and the measures necessary to secure adherence to the best practicable means. *Improvement notices* are now also sometimes employed. The Inspectorate's policy of reliance on extra-legal methods of persuasion and informed advice, rather than penal sanctions (fines are financially derisory), has a lengthy history[11]. It has tended to be generous in allowing existing works to continue to operate using controls much inferior to those specified in updated best practicable means.

The ultimate sanction, that of refusal of renewal of the certificate of registration, has never been employed in England and Wales, as the Chief Alkali Inspector stated in 1963:

> Annual renewal of the certificate could not be refused because a works did not continue to use the 'best practicable means'. On the contrary, the Inspectorate would be failing in its duty if it allowed such a thing to happen (Social Audit, 1974).

Similarly, no instance of refusal to grant a certificate of registration in the first place appears to have been reported. In other words, air pollution permits are always given though, in principle at least, the best practicable means for a particular works could be made so stringent as to be tantamount to a refusal.

The reluctance to take enforcement action (at least public enforcement action) is one of the characteristics of HMIP. Another is the lack of information available to the public about both the process of setting industry-wide best practicable means and their application to a particular works. While representatives of certain interested bodies have been consulted recently on best practicable means and local authority environmental health departments are now informed of the conditions applied to registered works in their areas, the public have not been involved or informed.

Criticisms of this secrecy were taken up by the Royal Commission on Environmental Pollution (1976) which recommended the release of emissions data, the publication of breaches of requirements and the general availability of registration conditions for particular works. There has since been a marginally greater release of information and the Inspectorate has encouraged the setting up of local liaison committees, at which pollution problems from particular works are discussed.

There is, all in all, remarkably little evidence on which to review the real achievements of HMIP. There are, for example, no annual estimates of air pollution control expenditure of the type published in the United States, and it is not possible to advance even rough totals of spending for comparative purposes. There appears to be little doubt that industrialists' anguished prophecies of insolvency if further pollution arrestment is demanded receive a more sympathetic hearing than the protests of those who suffer the effects of emission from registered works (Miller and Wood, 1983:17–27). The alacrity with which industrialists seek registration and the partiality of the advice of the Inspectorate to local planning agencies (Chapter 6) tend to indicate that HMIP might be in the control of the regulatees. On the other hand, the Royal Commission on Environmental Pollution (1976) was reluctantly impressed with the way control was achieved (pp 73–84). The Inspectorate itself has come to accept that there may be weaknesses in the best practicable means approach:

> The present system may not be perfect, but it has stood the test of time and, at a comparatively low administrative cost, has produced environmental conditions which bear comparison with those in the best of other industrialised countries (HSE, 1984:21).

Peacock (1984) has confirmed that, while evidence about the costs of air pollution regulation in Britain is very difficult to obtain, they appear to be very small. Hill (1983) has rejected the notion that HMIP is in the control of the regulatees, despite a consensus being generally shown to outside society:

> Every pollution control system involved interactions between interested parties, and strenuous efforts by the economic interests to minimise limitations upon their activities. The British system does at least make this very clear (p 105).

As Rhodes (1981:153) points out, it is curious that local authority environment health officers have largely escaped criticism, despite the fact that they appear to prosecute a smaller percentage of offenders against the Clean Air Acts than does the Inspectorate under the Alkali Act (RCEP, 1976). (The fines they levy are again small, seldom exceeding £100.) The absence of national information of the type provided in the Inspectorate's annual reports must be a contributing factor to public acquiescence. There are, for example, no figures available to estimate total national local government manpower devoted to air pollution control. The officers concerned with air pollution in local environmental health departments are typically members of very small groups and some of them may not be technically qualified. Others, of course, *are* qualified and some local authority environmental health departments have proved to be highly competent in controlling pollution, notwithstanding the limitations of the powers available to them.

The officers in environmental health departments have relatively little discretion in interpreting the provisions of the Clean Air Acts. If chimney heights, for example, are considered adequate when calculated according to a standard formula, permission must be granted. Their other, very limited, anticipatory powers are similarly rigidly prescribed. Providing a developer meets these requirements, a permit must be granted. Partly because of this lack of discretion, there has been very little political interest in air pollution control within local authorities and officers are normally left to make decisions without being influenced by local politicians. There is no other external input to the officers' decisions, since public participation is virtually non-existent. In practice, there is a good deal of activity by environmental health departments which is not directly sanctioned by the powers conferred by the Clean Air Acts. This partly manifests itself in the use of land use planning powers to prevent pollution. As the Commission on Energy and the Environment (1981) stated, with some hyperbole:

> Local authorities have no power to act in anticipation of a possible pollution problem – except in so far as they may be involved in giving or refusing planning permission for development (p 189).

Reform of the air pollution control system has been expected for a number of years and the Department of the Environment (1986) issued a consultation paper which proposed several changes. These were intended to retain the principle of control of industrial emissions through the use of best practicable means but to clarify its breadth of application and allow its use to be adapted to take account of existing and prospective European Community legislation, including that on standards and that embodied in the framework directive requiring 'no significant air pollution' from industrial plants. It was proposed that a common system of control would be put forward which would give local authorities powers of control (including the prior approval so obviously needed) over certain processes similar to those exercised by HMIP. A new system of consents which would set out the main elements of the best practicable means agreed by the control authorities was put forward. Monitoring and enforcement proposals were advanced and suggestions for improving the hitherto extremely limited public access to information about air pollution control were also made. Consents were no longer to be annual but would be legally binding and subject to a charge (DOE, 1986). At the time of writing, it appeared that there was general support for these reforms and every expectation of their becoming law.

LAND USE PLANNING CONTROL

Legal framework

There had been concern over the effects of unplanned urban growth in Great Britain as early as the 1840s. This led to the passing of public health legislation and of byelaws to improve environmental standards in the latter part of the nineteenth century. The first year in which the word *planning* appeared in an act of Parliament was 1909. Between then and the 1930s several acts were passed which permitted the numerous local authorities to implement planning schemes. These schemes, in many ways similar to the US zoning provisions (Clawson and Hall, 1973:160), were largely ineffectual. Partly as a result of the writings of pioneer planners like Geddes, Howard and Unwin, a series of commissions and inquiries was set up during the Second World War. The reports of these led directly to the reforms embodied in the Town and Country Planning Act 1947.

This act scrapped the previous permissive planning system and ensured that planning control be administered only by the largest units of local government and that these be responsible for the preparation of plans and for the control of development. The financial aspect of land development was covered by the nationalisation of development values (a logical consequence of the effective nationalisation of development entitlements), leaving owners only with the right to continue using their property for its existing purpose. A national fund was set up to provide the resources to compensate land owners who would then receive no further profit if the local planning authority granted permission for a change of use of land. This aspect of the planning legislation did not work satisfactorily and, despite several subsequent attempts to recoup for the community the benefits to private owners of increased values resulting from the actions of planning authorities, it no longer exists. Public acceptance of the principles of plan making and development control has continued largely undiminished.

In 1968 a new system of plans was introduced and, in 1974, local government reorganisation swept away the myriad small local authorities and replaced them with a two-tier system of counties and districts. The metropolitan counties were abolished in 1986 leaving planning powers with the shire counties and their constituent districts in the rural areas and with metropolitan districts in the conurbations. Various amendments to the planning system, including simplified planning zones (which follow the introduction some years ago of enterprise zones) have been made. In both types of zone it is possible to carry out various classes of development without requiring planning permission, providing it is in accord with an approved scheme (DOE, 1987a). Nevertheless, the land use control system remains essentially as it was conceived in the 1947 act: the preparation of plans and their implementation, at least partially, through control over development[12].

Development plans
Plans consist of a series of documents, including a written statement and maps or diagrams, containing a local planning authority's main objectives for land use in its area over a period of several years. *Structural plans* are prepared by the county planning authorities (except, from 1986, in the conurbations) and consist of a written statement, illustrated diagrammatically and setting out and justifying policies and general proposals for the development and other use of land in the area. These *must* include measures for the improvement of the physical environment, which definitely includes pollution control (Wood, 1979) and the management of traffic [13]. The policies and general proposals must be set in their general context – showing, for example, the implications of investment – and must indicate any *action areas* where comprehensive development, redevelopment or improvement is expected to start within ten years. There are elaborate provisions for public participation in the preparation of structure plans, including an examination in public – a type of public inquiry – before they can be approved by the Secretary of State for the Environment (in England and Wales).

Local plans, including plans for action areas must conform generally to the structure plan and are normally prepared by district planning authorities, although county planning authorities sometimes prepare certain types of local plan. They consist of a written statement and a map setting out the authority's proposals for the development and other use of land for the area, and define precisely the areas of land affected by the proposals. The plans deal with the detail of development, and so provide the basis for development control and for co-ordinating public and private development. They must include such measures as the planning authority thinks fit for the improvement of the physical environment, including the control of pollution[14]. This is a weaker requirement than in the case of structure plans.

Besides the plans for action areas, there are *district plans* for the comprehensive planning of relatively large areas, usually where change will take place in a piecemeal way over a long period, and *subject plans* for dealing with a particular type of development (such as the reclamation of derelict sites) in advance of the preparation of a comprehensive plan. Again, there are extensive provisions for public participation, including the holding of a public inquiry. The planning authority normally adopts the local plan following an independent report by the inspector adjudicating at the inquiry (Cullingworth, 1985). In the metropolitan areas, the districts are now required to prepare *unitary plans* for the whole of their areas, while *strategic guidance* from the Department of the Environment takes the place of structure plans.

Control of development
With certain limited exceptions all development (which includes most forms of construction, engineering, mining and any material change in the use of

land or existing buildings) requires the prior consent – planning permission – of the local planning authority. Nowadays, the shire counties deal only with minerals and waste disposal applications and a limited number of other such county, (that is, broader area strategic matters) leaving all other decisions to the districts.

When determining the application for planning permission, the authority must keep in mind the provisions of the development plan for the area concerned and any other relevant considerations – for example, the effect of the development on the ambient air pollution level in the area. If proposals for development do not accord with the plan, the local planning authority can still give its consent if it believes that they do not conflict with or prejudice a fundamental provision of the development plan.

Where departures from the plan do involve conflict with its fundamental provisions the authority must, if it proposes to permit the development, give public notice of the application and ask for representations, and must send a copy of the appplication to the Secretary of State for the Environment. The Secretary of State does not normally intervene unless it appears that important planning principles or issues of more than local significance are involved, but he has powers to call in the application to make his own decision, or to direct that planning permission be refused, or to leave it for decision by the local planning authority.

After considering an application for planning permission, the local planning authority can grant unconditional permission, refuse its consent, or grant its consent subject to conditions. Such conditions may require, for example, that the site be landscaped, or that a process be operated only during specified hours (DOE, 1985a). No compensation is paid for refusal of consent or for the imposition of conditions on a consent.

There is a right of appeal to the Secretary of State against refusals or against the conditions attached to a grant of permission. This right of appeal by the developer does not extend to third parties (see later). While public inquiries to hear appeals are often held, the majority of minor appeals are dealt with by written representations. The decision of the Secretary of State (or his inspector) following an appeal may only be challenged in the High Court on points of law, not on the merits of the case. Public inquiries frequently represent the only opportunity for public involvement in major land use and environmental decisions and they are consequently often broad-ranging. Pollution control, for example, is frequently discussed at length and in considerable technical detail in inquiries into industrial development proposals.

There are other provisions for public participation in development control decisions. Registers of all planning applications have to be kept and made available for public inspection, and applications for certain types of development – broadly those which might be regarded as anti-social on noise, odour or other nuisance grounds – must be advertised locally giving the general public the chance to object. Under the Local Government (Access to Information) Act 1985, the public has the right to attend local authority committee meetings, and to peruse background papers, reports and minutes of meetings. Many planning committees hear objectors in person. While these objections may be taken into account by the local planning authority before planning permission is granted, there is no opportunity for the public to object against its grant of planning permission, once issued.

There are special planning requirements in *conservation areas*, in *national parks* (which are not owned by the nation), and in *areas of outstanding natural beauty*. There is also legislation relating to the construction and operation of new towns. *Green belts* have been designated around most major cities, in which there is a strong presumption against development.

If development is carried out without permission (or conditions are not complied with), the local planning authority may serve an *enforcement notice* (against which there is a right of appeal to the Secretary of State) specifying the steps which it requires to be taken for the purpose of remedying the breach of planning control. The authority has a right to prosecute developers who fail to comply with an enforcement notice but fines are generally not punitive.

Among the many sections of the Town and Country Planning Act 1971 (as amended) are powers relating to the use of voluntary *agreements* between the developer and the local authority to achieve planning gains for the community. There are also provisions relating to the discontinuance of existing uses and to the modification of planning conditions to an existing permission, both on payment of compensation to the developer. Much of the detailed operation of the town and country planning system is prescribed by two statutory instruments: the General Development Order[15] and the Use Classes Order[16] (Cullingworth, 1985).

Amenity is a key concept in town and country planning and, while it is not defined in legal terms, it is apparent that it subsumes pollution, since the Use Classes Order refers to 'detriment to the amenity of that area by reason of noise, vibration, smell, fumes, smoke, soot, ash, dust or grit'. Local planning authorities have very wide scope for deciding whether to grant a planning application and for establishing conditions which control the siting and design of a development and the manner in which activities associated with it are to be carried out, often in consultation with the developer. As the Commission on Energy and the Environment (1981) stated:

> There are no specific standards of amenity or design against which local authorities have to assess planning applications. It is their responsibility to decide whether the proposed development is acceptable in the context of the locality where it is to take place and the general environmental standards prevailing at the time (p 22).

Practice and reform

The administration of the United Kingdom land use planning system by local authorities involves a large number of qualified planners. In addition, numerous firms of consultants are engaged on commissions relating to planning applications and appeals. Further, numerous planners are engaged as central government officials and inspectors. Some lawyers also specialise in practice related to the planning system. The Royal Town Planning Institute, the professional association for qualified planners, has some 14,000 members.

There has been a widespread consensus about the utility of the land use planning system in the United Kingdom since its introduction. The one major area of disagreement was not the removal of development rights but the question of whether gains in the value of land caused by the grant of planning permission, or by other planning decisions, should accrue to the landowner or to the community. Another problem has been the length of certain public inquiries into proposals for major and controversial developments like nuclear power stations. While the vast majority of inquiries are short (less than a week) one or two have lasted over a year.

The Conservative Government, elected in 1979, swept away taxes on development gains soon after taking office. More than any other administration, it has questioned whether planning controls are sometimes too restrictive. It has held that public sector expenditure (from which planning is largely financed) must be reduced and that much of the planning system may be regarded as dispensable regulation, though it has latterly softened this view. It accordingly instituted six main reforms of the planning system, as well as simplifying the division of planning responsibilities between districts and counties.

First, the Government introduced charges for planning applications. The local authority was thus to make the applicant pay for the costs incurred in deciding whether, in the interest of the community as a whole, his project was the right type of development in the right place: a controversial measure. Second, the Government extended the range of permitted development, thus allowing more minor activities to proceed without requiring planning permission. Third, the concept of enterprise zones was introduced. While there was considerable concern that this loss of anticipatory planning powers might lead to pollution problems, in practice the construction of speculative industrial units together with the retention of pollution control agency powers appears not to have caused great difficulties (Wood and Hooper, forthcoming).

Fourth, simplified planning zones were introduced (DOE, 1987a). These zones will probably be confined to areas that are badly in need of regeneration and, as in the enterprise zones, pollution control requirements will continue to apply. Fifth, several inner city development corporations (not dissimilar to new town development corporations) have been created to funnel central government money into urban regeneration projects. These assume district planning powers in their designated areas.

The net effect of these changes has been less severe than might have been expected from the political rhetoric (Her Majesty's Government, 1985). The sixth reform may prove far more radical, however. The Government swept away the metropolitan county councils from 1 April 1986, leaving the conurbations with no regional land use planning and environmental management overview. This was, by far, the most controversial measure and it is likely to prove a considerable setback to planning in the conurbations, to co-ordinated pollution control in general and to air pollution control in particular (Wood and Jenkins, 1985). One other reform, the introduction of environmental impact assessment as a result of European Commission legislation (see later), is likely to have a marked effect on air pollution control.

Controls over national parks, areas of outstanding natural beauty and conservation areas have been retained but the highly successful new towns programme has been abandoned and the new towns' assets are being sold. After much heated discussion, green belts have been retained. These continue to cause considerable pressures for development close to them.

Structure plans have now been prepared for the whole of England and Wales and many revisions of these have been approved by the Secretaries of State for the Environment and Wales. Numerous local plans have also come into being. Even where no formal local plan exists, many planning authorities have prepared *informal* plans (plans as yet unapproved by the council of the local government) to help them guide development. All local authorities employ qualified planners, mostly in specialist planning departments or sections.

Despite the existence of the various plans guiding development, there is still no certainty in the British planning system, notwithstanding the fact that development control provides the main means of implementing statutory plans. In practice, however, development control is not such an integral part of policy making and implementation as was originally intended. Thus, if an area is shown to be industrial on the land use plan, a planning application for industrial development could be rejected by the local planning authority if it was, for example, thought likely to generate excessive air pollution. However, the pressures of Government policy, employment generation and, to a lesser extent, rateable values have been such that this course of action is becoming increasingly unusual.

Similarly, an industrial development might be accepted in an area designated on the plan for some other use if the local planning authority found convincing arguments for this – and employment generation may be enough. Recent Government policy guidance states very clearly that there is a 'presumption in favour of allowing applications for development'; that the developer is not 'required to prove the case for the development he proposes to carry out'; and that 'development plans are therefore one, but only one, of the material considerations that must be taken into account in dealing with planning applications' (DOE, 1985b). The Government has made it clear that priority should be given to industrial development and that permissions should be granted for small-scale commercial or industrial activities wherever possible, while not neglecting:

> ... health and safety standards, noise, smell or other pollution problems ... Where there are planning objections it will often be possible to meet them to a sufficient degree by attaching conditions to the permission or by the use of agreements under Section 52 of the Town and Country Planning Act, 1971, rather than refusing the application. Such opportunities should be taken (DOE, 1980).

This emphasis on the use of planning conditions and agreements rather than refusal is clearly not intended to extend to conditions overlapping the controls of HMIP. Notwithstanding this advice, it is extremely unlikely that polluting industry would seek to or be permitted to locate in a national park, area of outstanding beauty, green belt or conservation area or indeed in the heart of a homogeneously residential area.

While there may be no certainty of gaining permission for industrial development, the probability of acceptance is high. About 85% of all planning applications are approved (over 90% for industrial developments) and, of those applications which are determined by the Secretary of State after appeal against the planning authority's decision, nearly 40% are now approved (more for industrial developments) (DOE, 1987c).

There is, of course, much informal consultation about applications for major developments before they are submitted and the relationships established tend to discourage applications which are very unlikely to succeed. There is much more consensus between the land use planning authority and the developer about the nature and order of the physical environment than in the USA. Despite the uncertainties, therefore, a developer has a good chance of obtaining planning permission to construct a new source of air pollution in the United Kingdom, depending to some extent on the type of pollution involved.

ENVIRONMENTAL IMPACT ASSESSMENT

The Commission of the European Communities became interested in environmental impact assessment in the early 1970s. Following a number of EIA research programmes, the Commission decided that an EIA system should meet two sets of objectives:

1 to ensure that distortion of competition and misallocation of resources within the EEC are avoided by harmonising controls;

2 to ensure that a common environmental policy is applied throughout the EEC.

Accordingly, the Commission issued its first preliminary draft directive in 1978. After 20 such drafts, not all of which were released, and substantial consultation – this is reliably reported to have been the most discussed European draft directive ever – the Commission finally agreed the directive in June 1985, after significantly weakening its provisions at the behest of the British Government (CEC, 1985).

The approved directive specifies that projects likely to have a significant effect on the environment are to be subject to an EIA. Such an assessment is to be obligatory for virtually all projects, other than modifications to existing installations, in certain specified categories listed in Annex 1. These are oil refineries, coal gasification and liquefaction plants, large power stations, radioactive waste disposal sites, integrated steel works, asbestos plants, integrated chemical plants, motorways, railways and large airports, ports and canals and toxic waste disposal facilities.

An EIA is also obligatory for projects in certain other specified categories, listed in Annex 2 of the Directive, and for substantial modifications to Annex 1 projects, but subject to criteria and thresholds to be established by member states. Annex 2 includes many industries not encompassed by Annex 1. In addition, an EIA is required for any other projects outside the above categories where a significant environmental impact is likely to occur.

The developer bears primary responsibility for supplying all the relevant basic information required in an environmental impact study. At the same time, it is envisaged that the competent authority, ie, the local planning authority in Britain, may often need to assist the developer in the preparation of the study. The authority also has the responsibility of checking the information supplied, which includes information similar to, but less comprehensive than, American requirements. Further specification of the desirable content of an assessment is provided in Annex 3 of the directive.

There are provisions for public and agency consultation. The competent authority has to publish the fact that the application had been made, make all the environmental documentation available to members of the public and make arrangements for concerned parties to present their views. The authority has then to make its decision and publish this, the reasons for granting or refusing permission and the conditions, if any, to be attached to the granting of the permission.

There has been considerable official interest in EIA in the UK for some years. Several reports have recommended the acceptance of an EIA system but governments of both parties have been very cautious in their attitude to it[17]. The net effect of the deletions of various provisions from earlier drafts has been the emasculation of original intentions, but the directive will still affect the operation of the UK planning system, since it seeks to apply the same provisions to public as to private developments and to provide for more public information prior to any siting decision, whether following a public inquiry or not. This last consequence is likely to prove the most far-reaching for practice as it will vastly increase the amount of information available to third parties outside the inquiry process. Regulations[18], guidance and indicative criteria for the inclusion of Annex II projects (DOE, 1988) in the UK have been issued. These require the publication of an *environmental statement* as a result of the *environmental assessment* of a proposed development.

Some 200 environmental impact assessment reports (not necessarily meeting all the European Commission's criteria) have already been carried out in the UK (Petts and Hills, 1982). Most local planning authorities undertaking these have been well pleased with the results and have not experienced untoward delays in determining applications. The costs (while difficult to determine) appear not to have been exorbitant, normally being less than about 0.5% of project costs (House of Lords, 1981). It seems likely, therefore, that the mandatory system of EIA could be extended informally to types of projects unspecified in the directive by authorities determined to consider carefully the possible impacts of potentially polluting developments.

Any environmental impact assessment of a potential pollution source is bound to include air pollution impacts. This will involve forecasting these impacts, their significance and considering mitigation measures as part of the planning system. The planning authority should consequently have the benefit of sufficient information on the likely air pollution effects, and of the opinions of the air pollution control authorities before the planning decision is taken, as the regulations require the authority to consult HMIP where an environmental assessment is undertaken (DOE, 1988). It should therefore be able to request appropriate control measures while the project is still at the design stage. EIA should, if it is implemented appropriately in the UK, lead to improvement in the consideration of air pollution in planning decisions.

NOTES

1 Office of Population Census and Surveys (1983). This is an underestimate of the number of coloured persons resident in Great Britain as it does not include those born in the country.
2 For a fascinating history of the control of air pollution in the United Kingdom, see Ashby and Anderson (1981).
3 The Health and Safety at Work Act 1974, and the Control of Pollution Act, 1974, repealed the majority of the lesser provisions of the Alkali Act. It is proposed to repeal the Alkali Act in toto eventually, but to perpetuate similar powers of control (ie, the best practicable means approach) by other means.
4 *Alkali, etc, Works Regulation Act 1906* s 7(1).
5 *Clean Air Act 1956* s 34(1).
6 The presumptive limits are published in the Inspectorate's annual reports.

7 These notes are now listed in appendices to the Inspectorate's annual reports. Several are reproduced as an appendix in McLoughlin and Foster (1982).
8 *Clean Air Act 1956* s 10. This right of approval applies to any chimney emitting gases and not merely gaseous products of combustion.
9 The Secretary of State for the Environment can impose general limits on the sulphur content of fuels.
10 There were 7 vacancies in the nominal establishment of 44 at the end of 1986 (HSE, 1987).
11 Only two prosecutions appear to have been carried out in 1986 (HSE, 1987).
12 See, for a good exposition, Cullingworth (1985).
13 *Town and Country Planning Act 1971* s7(1)A(a) (as amended).
14 *Town and Country Planning Act 1971* s11(1)a (as amended).
15 *The Town and Country Planning General Development Order* 1988, SI 1988, No.1813.
16 *Town and Country Planning (Use Classes) Order 1987* SI 1987 No. 764.
17 For an account of the gestation of EIA in the UK, see Wood (1982). See also House of Lords (1981).
18 There are over a dozen sets of regulations. Those relating to the English planning system are: *The Town and Country Planning (Assessment of Environmental Effects) Regulations 1988* SI 1988 No. 1199.

6

Implementation of Preventive Controls in the United Kingdom

This chapter parallels Chapter 4. It illustrates how air pollution controls and land use controls affect the siting of new or modified stationary sources of air pollution. The relationship between land use and air pollution controls in practice is first discussed. Eight case studies of the implementation of these controls are then presented.[1]

RELATIONSHIP BETWEEN LAND USE AND AIR POLLUTION CONTROLS

In the absence of realistic formal anticipatory pollution control powers over non-registered sources, UK local authorities are bound to try to employ land use planning controls to reduce potential pollution problems. The planning departments of local authorities have shown a readiness to use statutory powers, both in the preparation of development plans and in the control of development, to prevent atmospheric pollution or to mitigate its effects upon the population.

There is positive encouragement from central government to include air pollution policies in development plans:

> Both structure and local plans are required to include land use policies and proposals for the improvement of the physical environment. This includes ... policies designed to control pollution and to limit and reduce nuisances such as noise, smell and dirt. In formulating all their policies and proposals, local planning authorities should have regard to the impact the policies and proposals will have on the environment and how they relate to pollution control; this should be made clear in the explanatory memorandum or reasoned justification. In particular the introduction of European air quality standards may impose a constraint on the extent to which plans should provide for intensification of development in some urban areas (DOE, 1984).

While encouraged, it is not proposed to make the inclusion of such air pollution policies in plans mandatory (Central Directorate on Environmental Pollution, 1982:16).

The coming into effect of the European Commission's air quality standards was a very significant event in the control of air pollution in Britain: it implied a change from case-by-case control to at least some recognition of the effects of air pollution on and from surrounding areas. Notwithstanding the European Commission's annual standards for sulphur dioxide and smoke, the Government did not modify the tenor of its advice to local authorities. Local authorities containing areas where limits were exceeded were merely urged, in the Government's circular, to 'take into account the need to attain the limit within a reasonable time' (DOE, 1981). There was no question of the prohibition of development in such areas. However, there are signs that this attitude is changing, as mentioned in Chapter 5.

Many structure plans demonstrate a real grasp of air pollution problems, a recognition of the use of planning powers in the control of pollution, an appreciation of the roles of land use planning in pollution control and an awareness of the various planning techniques for controlling pollution (Wood, 1979). There is considerable variety in the treatment of air pollution in local plans. This ranges from the detailed to the non-existent. There is, of course, no statutory requirement to include 'measures for the improvement of the physical environment' in local plans unless the planning authority thinks fit (Chapter 5). There appear to have been no subject plans relating to air pollution.

Development control powers have been found, of course, to be most effective when applied to new (as distinct from established) development where they can be used to prevent the juxtaposition of the polluter and land uses particularly susceptible to pollution damage. Few British planners possess any detailed knowledge of pollution control technology and it is generally accepted that planning powers should be used to ameliorate air pollution problems only after consultation with the pollution control authorities: HMIP and the environmental health departments. In the latter case, consultation usually consists of discussions between the officers of a single district council. In the case of liaison between planning departments and the Inspectorate, relations have occasionally been strained by antagonisms and conflicts of interest which stem from fundamental differences, in both attitudes and responsibilities, between the local and the central bodies.

The Inspectorate is now asked for its advice on planning applications for registrable works by most local planning authorities as a matter of course. In some cases, the advice received would appear to be perfunctory, consisting of a simple reiteration of the operator's obligation to comply with the best practicable means without specifying what they will consist of, or more pertinently, what the consequences of the permitted discharges might subsequently be. The consultative role of the Inspectorate is not confined to offering advice on the location of potential sources of atmospheric pollution. It is willing, indeed often eager, to comment on the wisdom of permitting sensitive development, notably housing, in the vicinity of existing registered works:

> The Royal Commission on Environmental Pollution recommends there should be a mandatory obligation on planning authorities to consult the Inspectorate on all applications to build or alter registrable works. The Health and Safety Commission support this recommendation and consider that the obligations should extend to consulting about other developments near to existing registered works (HSE, 1978:3).

While agreeing with the concept of consultation, however, the Government chose not to make this compulsory (Central Directorate on Environmental Pollution, 1982).

It is apparent that HMIP's view of planning controls tends to be limited to controls over development around a works but seldom, if ever, to a recommendation not to construct a works. As the Deputy Chief Inspector stated:

> We get very good consultation with planning officers ... Inspectors ... are often asked a question which they can't really answer: 'what effect is this plant going to have upon the environment?' Quite often ... all we can say is that, in our honest opinion ... you should or should not permit the neighbouring development to come nearer than so many metres or kilometres to the works (Miller et al, 1980:63,64).

Again:

> A district Alkali Inspector gave evidence at a Planning Inquiry where a company appealed against refusal by a local planning authority, quite rightly in our view, to grant permission for housing development alongside a mineral works (DOE, 1975:10).

Other examples of this attitude can be quoted (Miller and Wood, 1983).

Notwithstanding this seeming partiality, the necessity to prevent new industrial development where pollution may damage the environment is recognised by the Inspectorate, at least in principle:

> It is encouraging to see that an increasing number of planners are seeking the (Inspectorate's) advice on the air pollution aspects of new applications concerning both registered and non-registrable works.... But many are still failing to do so with the result that industrial developments which cause pollution are allowed too near houses, shops, schools or hospitals and vice versa (HSE, 1978:3).

It appears, however, that not only is best practicable means adjusted only marginally to take account of local circumstances, by altering chimney heights in certain cases, but that it is extremely unusual for HMIP to advise against the location of a new source of air pollution, whatever its neighbours (Miller and Wood, 1983). Though negotiations between the Inspectorate and the developer may be privately protracted, the requisite authorisation has always been forthcoming. It remains the function of the local planning authority to determine whether or not best practicable means is enough in the local conditions and circumstances, and hence to decide whether or not to permit the development of the new source.

The Inspectorate has uniformly resisted the use of planning conditions which might duplicate its own powers. It has been supported in this view by the Royal Commission on Environmental Pollution (1976) and by the Government. However, the practice of imposing planning conditions on new industrial premises not controlled by HMIP in lieu of other anticipatory powers, is acceptable to the government:

> where there are no ... such specific controls, it will continue to be appropriate to consider the use of planning controls to ensure that, where necessary, new development incorporates features which will make it acceptable from a pollution point of view in the proposed location[2].

Despite the existence of powers under the Clean Air Acts over smoke, chimney heights and dust and grit emissions, it would appear that it is quite common for local planning authorities to control such pollution sources at the request of environmental health departments. Similarly, and again usually at the behest of the environmental health departments, many local planning authorities have used planning conditions and planning agreements to secure some form of anticipatory control over premises which fall outside the scope of both the Clean Air Acts and the Alkali Act and have sometimes achieved this where premises fell within the scope of the latter legislation (Miller and Wood, 1983).

Notwithstanding the advice by central government on the importance of not promulgating plans which might lead to the exceeding of European air pollution standards it has stated that the Commission's air quality directive should 'not be interpreted as prohibiting the siting in such areas of new plants that may be sources of smoke or sulphur dioxide'. Similarly, as mentioned in Chapter 1, the sanction of planning refusal was to be used only when the development would result in a 'significant deterioration' of local air quality even after the use of specific powers to control pollution (DOE, 1981). It is perhaps a telling reflection on the effectiveness of those specific powers that despite their application, significant deterioration of air quality remains a possibility. In these circumstances it seems likely that planning authorities will continue to use their own statutory planning powers to prevent additional pollution of the local atmosphere, as several of the following case studies illustrate. Some of their characteristics are presented in summary form in Table 7.

122 *Planning Pollution Prevention*

TABLE 7 Some characteristics of the UK case study siting processes

	Air Pollution Control Agency			Land Use Planning Agency					Other procedures			Implementation				Use of the Courts
	Number of agencies	Prior objections	Accept/Refuse	Appeal (Acc/Ref)	Number of agencies	Prior objections	Accept/Refuse	Appeal (Acc/Ref)	Air pollution control conditions	Prior objections	Accept/refuse	Construction	Subsequent objections	APCA enforcement	LUPA enforcement	
New Town	1	-	-		1		-		-			-	-	-	-	
Tameside	1	-	-		1	-	-	X	-							
Bolton: Incinerator	1	-	-		1	-	X	X	-							
Yorkshire	2	-	-		1		,;X	-	-			-	-	-	-	
Bolton: Chloride	2	-	-		1	-	-		-			-	-	-	-	
Cheshire	1	-	-		2	-	(,)	-	,;X			-	-	-	-	
St. Helens	1	-	-		1	-	-	(,)	-			-	-	-	-	-
Glossop	1	-	-		1	-	-	-	-			-	-	-	-	(,)

THE NEW TOWN GLASS FIBRE WORKS

A company wishing to manufacture glass fibre approached a new town development corporation in 1977 with a view to constructing a large factory in an area zoned for industrial use[3]. This zoning, which must be approved by the Department of the Environment under the provisions of the New Towns Act 1965, is made without any knowledge of the type of industry likely to be attracted and without any necessity to carry out consultations. Once designated, incoming industries do not need to apply for planning permission but a lease is drawn up between the development corporation and the company, to which covenants may be added. It is thus similar to enterprise zone and simplified planning zone arrangements. Development corporations do not usually employ staff with pollution control expertise and are under no duty to consult other bodies in drawing up the lease. The nominated members of the corporations are not normally involved in such discussions.

The company concerned was extremely reticent in providing information about its proposed process, which it felt to be commercially sensitive. The development corporation nevertheless requested more details and it was apparent from these that potential air pollution problems would arise. The only site suitable for the development lay in an industrial zone close to the boundary of the new town and also close to housing, a hospital, and food and drink manufacturing plants. The new town corporation was anxious to attract the development but was now conscious that an independent assessment of the environmental consequences was necessary. Accordingly, it gave its agreement in principle and commissioned consultants to investigate and to recommend conditions which could be incorporated into the lease as covenants.

A few weeks later, the environmental health department of the district council in which the new town was situated was belatedly consulted on noise aspects. It immediately took a wider interest in the proposal. The consultants recommended a chimney to reduce air pollution, but the environmental health department expressed its concern that the proposed chimney was of inadequate height to render ground level concentrations acceptable and requested further details of stack emissions. HMIP was consulted, though glass fibre manufacture is not a registered process, and voiced its concern about likely air pollution and detrimental effects on the adjoining food and drink processing plants. In 1978 the consultants suggested a higher chimney height of 24m but the environmental health department, in consultation with HMIP, felt that 46m would be more appropriate though it still believed it had inadequate information.

The development was now discussed by the district council environmental health committee, meeting in private. It recommended the incorporation of a set of stringent pollution control conditions in the company's lease and the use of independent consultants to make an assessment of air pollution problems. It hoped this would break the deadlock which had developed between the environmental health department on the one hand and the company, the development corporation and their consultants on the other.

The development corporation felt that some of the suggested conditions were too specific but the district council continued to press for the incorporation of air pollution controls. The council duly appointed consultants, who suggested a chimney height of 55m and permission for this was granted, under the provisions of the Clean Air Acts 1956 and 1968. Several stringent conditions relating to emission limitations and to the process were appended, some of which appeared to be not strictly within the scope of the acts. The lease, with many air pollution control conditions, was eventually signed in 1980 and construction was completed in 1981.

A number of complaints about phenolic odours, eye irritation and particulate emissions from local residents, who had not been used to industrial neighbours, ensued. Measurements taken by the environmental health department demonstrated that the plant was meeting the various emission limitations imposed upon it and none of the pollutants could be detected in the atmosphere using monitoring equipment. Alterations to the process were made by the company at the request of the environmental health department but these led to increased emissions of glass fibre resin and renewed complaints. Further, quite costly, changes to the gas-scrubber were made which led to some improvement but, at the time of writing, the chimney appeared still not to be functioning properly and the council was contemplating insisting upon the use of incineration. The level of complaint about organic odours and about irritation from a fine blue smoke was high and local residents were expressing deep disquiet and suggesting the formation of an action group. The council had not resorted to legal proceedings because the company was perceived to be co-operating in trying to reach a satisfactory solution.

The district council's involvement in the siting decision was incidental although it (and not the development corporation) was statutorily responsible for pollution control. This illustrates the importance of meaningful consultations in the siting of a new source. A subsequent circular has urged development corporations to consult other bodies in cases like this. Only controls by means of conditions could be imposed because the development had already been agreed in principle, but the district council was able to demand a high chimney and to impose numerical emission conditions on the chimney height approval, its only statutory power. The new town development corporation also included pollution control covenants in the lease which were more enforceable than normal planning conditions. When problems arose, however, the new town development corporation left it to the district environmental health department to implement better controls. Despite the fact that the various conditions were not exceeded and despite the construction of the high chimney, the vigorous complaints of the public ensured that action by the air pollution control agency and the developer was taken to try to achieve a solution.

THE TAMESIDE RESIN MANUFACTURING MILL

Sterling Mouldings Ltd applied for outline planning permission to extend its existing premises (an old cotton mill) to enable a resin manufacturing process to be carried out on site in 1976[4]. The mill is situated in Tameside, Greater Manchester and is virtually surrounded by modern housing, much of which results from redevelopment of the area. The existing resin moulding powder processes carried on at the mill had given rise to a history of complaints, and the area around another of the company's nearby works had had to be evacuated on several occasions because of releases of toxic chemicals. The application contained very limited information but it was stated that a number of hazardous chemicals would be stored in quantity, including formalin and phenol: both to a maximum of 120 tonnes. The residents of the new houses vociferously attacked the proposal because of the potential pollution and because of the reversal of the trend towards residential use in the area. They enlisted the support of their three local councillors.

Numerous consultations were carried out by the local planning authority on the basis of the information supplied. The environmental health department gave assurances that 'no unacceptable pollution should result from the process' and that safeguards relating to fume and dust emissions could be attached to the detailed permission. Two public meetings were held at which the proposals were explained. Heated exchanges between the public and officers took place but the previous record of the company was ruled to be irrelevant to the present decision. While the officers' report to the planning committee mentioned existing pollution, the environmental health officer gave many assurances regarding the safety of the plant and dismissed residents' fears. The committee (not the full council) granted the permission, after lengthy discussion, in 1976, subject to a number of conditions, including one intended to prevent any possible pollution.

After pressure from the three local councillors, and the mention of new evidence, the full council asked the planning committee to reconsider its decision, knowing that the withholding of permission would now involve a revocation order and the payment of compensation. When the planning committee met again, it heard evidence from the local residents, the council's officers and the company and decided not to revoke the permission, but only on the chairman's casting vote.

Before the full council next met, at the end of 1976, the residents obtained scientific evidence in support of their objections from the regional branch of the British Society for Social Responsibility in Science. This referred to the potential of phenol and other chemicals for causing damage by fire or explosion when stored in bulk. At the council meeting, one leading councillor dismissed this evidence as 'a load of emotional clap-trap provided by students' and another claimed to have gargled with phenol. After hearing further representations, the council endorsed the planning permission, but again only on the casting vote of the chairman. The Local Government Ombudsman, who investigated the case, found no maladministration, but felt the initial decision should have been taken by the full council, and not the committee, because of its controversial nature.

The application for reserved matters was made at the end of 1977, and a new officer in the environmental health department requested further information on various pollution aspects and subsequently recommended pollution control conditions. Permission was granted in 1978, with conditions. Later that year another of the company's mills was evacuated when a fire occurred and the company was prosecuted by the water authority for releasing 500 gallons of styrene into a water course. These incidents were given wide publicity and, at the end of the year, following further pressure from the local councillors, the council voted to revoke planning permission, only to reverse this decision in 1979 after press speculation that a compensation claim of £3 million might be involved. In the event, the company decided to concentrate the manufacturing of resins at another site, and continue to make only powders at this mill, largely as a result of the widespread public opposition.

The case demonstrates the importance of the political context in land use planning agency actions, the attitude of air pollution control agency officials and the reputation of the developer most clearly. The campaign by the three ward councillors, which was so nearly successful in convincing the council to revoke planning permission, was matched by the ignorance of other elected representatives. The initial report by the environmental health officer that there would be no problems was unlikely to be convincing in the light of experience from the developer's previous activities on the site and at other works in the locality. The strength of public opposition was reflected in the unusual permission to allow members of the public to address the planning committee and in the developer's eventual change of mind about implementing the project. However, objectors had insufficient information and time to mount an effective argument before the crucial initial permission was given.

THE BOLTON SEWAGE SLUDGE INCINERATOR

The North West Water Authority applied in 1976 to Bolton Metropolitan Borough Council for outline planning permission to build a sewage sludge incinerator on the site of an existing sewage works located in the Irwell Valley. The site was designated as a 'green area' in an informal plan prepared some years earlier. Substantial derelict land reclamation and planting schemes had been implemented in the area. The water authority intended to bring sludge from a variety of locations for incineration as an alternative to the method of disposal then currently employed – dumping at sea, via the Manchester Ship Canal.

It was agreed between Bolton and the Greater Manchester Council that the planning decision was a county matter, since strategic issues were involved. It would therefore be determined by Greater Manchester Council whose officers felt that the potential pollution problems arising from the development, apparent from the documents accompanying the application, were outweighed by the strategic sludge disposal requirements of the water authority. Bolton would, however, make a recommendation.

Bolton's planning department officers assumed a co-ordinating role and relied upon the environmental health department to supply the necessary expertise on air and noise pollution, while dealing with other planning considerations themselves. They consulted widely about the application in order to prepare their recommendations. Through the planning department, environmental health officers asked the water authority to prepare an environmental impact assessment but the North West Water Authority declined, stating that it was pointless until planning permission had been secured! The only disagreement between the two district departments concerned the preferred height of the stack, the planners wanting the height limited to 50m for visual amenity reasons and the environmental health officers requiring a higher chimney to dispose of airborne effluents effectively and prevent the build-up of ground level concentrations in the Irwell Valley.

An existing residents' association was activated and operated with great professionalism, calling on the skills of a socially mixed population, marshalling arguments, obtaining expert advice, lobbying support in opposing the proposal from councillors and from their Member of Parliament, and gaining publicity in the local press.

As a consequence, despite a balance of opinion in favour of the scheme among the Bolton officers (provided stringent conditions were attached to the planning approval to minimise pollution) the elected members of the whole council did not just recommend refusal of permission to the county but exercised their right under the Local Government Act 1972 and refused permission on pollution, and various other, grounds [5]. The councillors were following the precedent set by many previous planning decisions for the area which had been made to preserve and enhance its open character in accordance with the informal plan.

The water authority appealed to the Secretary of State for the Environment against the refusal and a public inquiry was held in 1977, at which air pollution was a major issue, being discussed at length by various expert witnesses, with the North West Water Authority making significant amendments to reduce the environmental effects of its original proposals. Bolton's officers and the residents' association were satisfied that their efforts had been

128 *Planning Pollution Prevention*

justified by these modifications and they anticipated that the appeal would be allowed, subject to the measures discussed at the inquiry to ameliorate pollution and other environmental impacts. However, notwithstanding these mitigation measures and the water authority's sludge disposal needs, the Secretary of State upheld his inspector's recommendation that the appeal be dismissed on the grounds that the choice of site was 'environmentally unacceptable'. The inspector considered that the visual improvements achieved in the area, and the provisions of the informal plan, would have been nullified by so industrial a development. Air pollution, despite the discussion devoted to it at the inquiry, hardly figured in the reasons stated for the decision.

The case demonstrates the dependence of the air pollution agency on the land use planning agency for anticipatory controls in Britain. The environmental health department would have had to accept emissions from the incinerator, had permission been granted. Notwithstanding the existence of the informal plan, the first instinct of the professional planners had been to approve the development. They were able to utilise a close and effective collaboration with environmental health officers though the unresolved chimney height issue was virtually a textbook example of the differing interests of the two agencies. The role of the objectors was very important, as they organised rapidly and effectively and persuaded the elected representatives to overturn their officers' recommendations and refuse the development. The concessions made by the developer at the subsequent public inquiry illustrated the value of meaningful prior examination of proposals to construct new air pollution sources. Finally, the importance of the policy context in land use planning decisions was demonstrated by the inspector's decision adhering to the provisions of the informal local plan.

THE YORKSHIRE CHEMICAL FORMULATION PLANT

In 1976 a small rapidly growing chemical firm, Crewe Chemicals Ltd, bought an old rug mill in a valley site in Kirklees, West Yorkshire, some distance from the nearest housing[6]. It applied for planning permission for a change of use from rug manufacture to the manufacture, processing and packaging of chemicals including herbicides and pesticides, but provided little information. Kirklees Borough Council Planning Department, working closely with the environmental health department, undertook a number of consultations, and asked for advice about likely pollution from HMIP (although the works was unregistered). None of the consultees advanced compelling objections and planning permission was granted subject to a number of conditions, many of which related to pollution control. There was little public involvement in the decision.

Shortly afterwards, the company applied for permission to use land surrounding the works for parking and the storage of chemicals. Again, consultations were carried out and, again, numerous conditions were applied to the consent. However, neither of the sets of conditions was operative, because the company had not yet started to operate processes which took the works outside the category of a general industrial building as defined in the Use Classes Order. Hence no statutory change of use had taken place and planning permission was not needed for the new manufacturing activity.

When eventually in operation, however, the chemical formulation processes attracted considerable opposition and complaint because of defoliation of vegetation. (The local recreation ground lost its grass.) Monitoring by Kirklees failed to link this defoliation with the company and it proved impossible to detect the chemicals manufactured or used at the works in the environment. A local action group and a pharmaceutical company downwind of the works were by no means convinced by these measurements.

A further application, for the construction of a new building, was submitted in 1977 and aroused a welter of protest, including demands by the action group for an environmental impact assessment. This time, the local authority decided to refuse the application because:

> The district planning authority is satisfied that the proposed development would result in an unacceptable increase in environmental pollution, and, furthermore, that the environment could not be satisfactorily protected by the imposition of conditions (Miller and Wood, 1983:121).

A subsequent application, late in 1977, for the construction of storage tanks, which had already been erected, was also turned down on the grounds that it would represent an intensification of use and could lead to environmental pollution. The council considered taking enforcement action. An application for a loading bay was, however, granted in 1978 subject to numerous conditions, mostly relating to pollution control and not confined only to the loading bay. Other applications ensued. That for a fence was granted but those that would increase capacity, including a second application for the storage tanks, were refused.

In early 1979, after continuing public protest, the local authority served an enforcement notice on Crewe Chemicals, demanding the removal of the tanks. The company appealed, not against this notice but against the refusal of planning permission for the tanks. Meanwhile, the company issued a public statement which, in effect, apologised for the environmental shortcomings experienced while establishing its commercial viability and promised to do better in future. Later that year the presence of atrazine, a herbicide manufactured by Crewe Chemicals, was found in the air intake filter of the pharmaceutical company.

The inspector at the public inquiry in 1980 found that planning permission was not required for the tanks. The plant was registered under the Alkali Act later in 1980 (to the pleasure of the works manager) and, as a consequence, a change of use from general industry was involved and the original planning conditions now became operative. Crewe Chemicals became bankrupt but a new company, Pennine Chemical Services Ltd, operated with the same general manager. Kirklees actively sought observance of the conditions on the various planning permissions and took enforcement action, though this was complicated by the change of company ownership.

Since the change of ownership there have been fewer pollution incidents and none involving atrazine, the throughput of which has diminished markedly. Although it appears that environmental control is given higher priority by Pennine than by Crewe Chemicals, problems remain. Output from other types of process had increased and use of the site had intensified significantly at the time of writing. The company had been prosecuted by the Yorkshire Water Authority for unauthorised discharge of trade waste to sewer. The level of complaint and political concern about noise, about organic chemical odours, about external chemical storage and about hours of operation remained substantial and formal enforcement action under planning or public health legislation seemed probable.

In the event, the formulation plant has proved a regrettable neighbour. The original permission is now regarded by the local planning authority as unfortunate, despite its inclusion of various pollution control conditions attached after comprehensive consultations. This case perhaps illustrates the importance of the insensitivity of anticipatory planning powers most clearly, since no planning permission was necessary for the change of use from rug manufacture to chemical formulation (and hence no conditions applied). The difficulties of proving what was quite obvious to local objectors, namely that the herbicide formulated by the company was causing the defoliation, were also apparent. Both the planning department and the environmental health department of the local authority attempted to implement controls, but their enforcement activities proved ineffective. The central government decision in favour of the developer following the inquiry particularly surprised the planning officers, who felt they had a strong case for enforcement. The welcome given by the works manager to the change of pollution control agency was notable: he felt that there would be less interference with his activities from HMIP than from the district. Even when the management objectives governing the operation of the works changed with the change of its ownership, and some improvements in pollution levels started to take place, other problems surfaced.

THE BOLTON LEAD BATTERY PLANT

Chloride, a company manufacturing lead batteries, sought to expand production in 1973 by constructing a new factory on poor quality farmland adjacent to derelict colliery and open cast coal sites at Over Hulton, Bolton[7]. More than 400 new jobs were involved and Chloride was anxious to locate close to its existing works in Salford. The only dwellings within 400m of the proposed building were a dozen houses alongside a major road. The planning department of Bolton County Borough Council embarked upon an extensive series of consultations involving advice on air pollution and other matters. The environmental health department and the HMIP were both involved at this stage.

The medical officer of health of the local authority and the Department of Health and Social Security in London stated that they could see no health grounds for refusing permission but the adjoining authority, Worsley Urban District Council (the proposed development being close to its boundary), and local residents, expressed fierce opposition to the development. However, in 1974, Bolton's planning committee granted outline planning permission subject to several stringent planning conditions which overcame many of the environmental objections to the development, including Worsley's. One of the conditions related to the reserved approval of 'processes to be carried out; together with the precautions to be taken to avoid any form of pollution or risk of explosion' (Miller and Wood, 1983:194). The factory was to be located behind a mounded, planted buffer strip of open land.

Public antagonism to the development persisted, partly as a consequence of the debate then raging about the environmental effects of lead. This concern may partly have motivated the officers of the environmental health department to impose very onerous controls on lead emissions, some of which would be the local authority's responsibility and some HMIP's. Negotiations progressed; local government reorganisation intervened (with the newly created Greater Manchester Council suggesting alternative locations for the works – all of which Bolton resisted), and the company was asked to reduce the number of stacks and to filter emissions from every chimney to reduce the quantities of lead released to the environment.

When the application for detailed planning permission was made in 1975 another round of consultations was embarked upon, on the basis of a set of provisional planning conditions. The quantitative condition relating to the maximum rate of lead emissions was much tighter than HMIP's presumptive limit and the district inspector protested vigorously both about this and about the duplication of his powers. The idea of retaining consultants to advise on environmental pollution was floated within Bolton Metropolitan Borough Council but categorically rejected by the environmental health department.

To circumvent HMIP's objections, Bolton decided to try to achieve the desired controls by using a planning agreement, rather than by means of planning conditions. Chloride concurred as it was anxious both to obtain planning permission quickly and to be seen to be doing all it could to protect the environment. Permission was granted in 1975 and an agreement was drawn up with stringent air pollution conditions. This was signed in 1977. The company itself suggested a maximum lead emission of just under 4kg per week, at least an order of magnitude below that normally required by HMIP. However, the Inspectorate stiffened its requirement in this case to a level compatible with the local authority's. The factory, now in operation, is considered a show-piece of pollution control and for some time emitted far less than 4kg of lead per week as monitored at the stack.

However, a complete change of senior management shortly after manufacturing commenced resulted in lack of attention to maintenance schedules. Although no increase in ambient levels was detected, the readings from the stack emission monitor showed Bolton environmental health department that a significant rise had occurred. The local authority encouraged Chloride to reinstitute regular maintenance and performance is now once again below the limit set. The planning agreement was revised and the emissions standard weakened to an annual average level of under 4kg a week when Chloride introduced a reaction pot to make lead oxide alongside the existing mill. Ambient lead levels at the nearby houses have fallen and very few complaints about air pollution have been received since the company built on the site. Annual liaison meetings between the pollution control authorities and the company are held and the works is now used by HMIP as an example of what can be achieved. Chloride has subsequently earned fees by advising other companies about pollution control techniques.

Needless to say, this desirable result could not have been attained without the willing co-operation of the company, whose leading negotiator was serving as the President of the Institute of Occupational Hygiene while the environmental controls were being decided. The importance of consultations was also apparent. Because of the benefits to the local community, there was never any question of a refusal of planning permission in this case. The company still willingly made a large number of environmental concessions and substantially modified its design. The conflict between the local land use planning agency and the central government air pollution control agency over the use of anticipatory powers was marked and resolved in the local authority's favour by the use of a legal agreement over which central government could exert no influence. The use of a planted buffer zone as a planning technique for air pollution control was notable, as was the location away from existing homes. Despite the political nature of the pollutant involved, lead, the opposition from local residents and neighbouring authorities was not vociferous enough to bring sufficient pressure to bear on the elected representatives to resist the application. This may partly have been due to the company's good local reputation. The opposition may, however, have encouraged the council to obtain the maximum concessions from Chloride and to sidestep HMIP's objections. The attitude of the developer is very clearly marked in this case, both in the initial negotiations and in the subsequent performance of the plant. The case also illustrates the reluctance of HMIP to adapt its best practicable means to allow clearly attainable emission limits to be imposed in particular local circumstances.

THE CHESHIRE FERTILISER PLANT EXTENSION

A fertiliser plant was constructed by Shellstar Ltd on the marshes to the south of the Mersey estuary, close to other works and some distance from housing, in 1967[8]. This plant occasioned numerous complaints from the residents of the two attractive nearby towns of Helsby and Frodsham about odours and particulates over several years. There was proven damage to hedgerow vegetation and claims of damage to crops. The pollution arising from the Shellstar works thus became notorious. In 1973 the company sold its interest to UKF Ltd. The new owners applied for planning permission to extend the works in 1974 and the Secretary of State for the Environment decided to call in the application because of its controversial nature.

The works is located in the area administered by Ellesmere Port and Neston Borough Council, which was uncertain how to react to the application. Two adjoining district councils, Chester and Vale Royal, opposed the development. Cheshire County Council carried out consultations and engaged the services of a firm of consultants, Cremer and Warner, to provide advice to enable it to determine its stance at the forthcoming public inquiry on what would have been a county matter had not the minister intervened. Cremer and Warner saw no reason to prohibit the development but recommended a number of planning conditions, should the Secretary of State ultimately decide to approve the application. Local residents, however, were extremely active in lobbying their elected representatives to seek refusal of the extension.

The council of Ellesmere Port and Neston, as a result of this pressure, overrode its officers' advice that permission be granted but that a planning agreement be made to control pollution problems, and recommended refusal. More crucially, the elected representatives of Cheshire County Council's Planning Committee also ignored their officers' advice to grant permission and expressed their opposition despite their stated desire to attract industrial development.

HMIP was involved in negotiations both with the company, and between the company and the county's planning officers, from an early stage. It warned against Cheshire seeking to duplicate HMIP's statutory powers over air pollution by the use of planning conditions. The attitudes of the elected representatives of both the county and the district softened somewhat as the new company began to demonstrate its competence in running the existing plant, substantially reducing pollution. During negotiations, UKF also promised improvements in environmental conditions provided its extension were to be granted. On the basis of Cremer and Warner's suggestions, the county planners were able to negotiate planning conditions with the company which would be applied if planning permission was granted. Many of these related to air pollution control. No agreement could be reached, however, on a condition originally suggested by the consultants requiring the monitoring of gaseous and particulate emissions: the company argued that the monitoring condition would be too expensive to implement. The elected representatives of both Ellesmere Port and Cheshire eventually agreed to support the company's proposals but only subject to the most stringent safeguards in the form of the proposed conditions.

The public inquiry held in 1976 was notable for the articulate and reasoned contribution of members of the public, in the form of local amenity societies and parish councils and for the insistence by the planning authorities on adequate environmental controls, should permission be forthcoming. The district pollution inspector appeared on behalf of the company. Cheshire argued that its monitoring condition and other proposals did not duplicate HMIP's powers. The company emphasised its contribution to the balance of payments and to local employment and used the tacit threat of withdrawal. It was supported by the Transport and General Workers' Union. The inquiry inspector recommended that the application be granted, subject to the conditions agreed between the planning authorities and UKF to control noise and air pollution and to a condition 'requiring the monitoring of chemical emissions from the plant'. He therefore took the county's part against the company on the one area of disagreement between them.

The Secretary of State's decision to grant conditional planning permission was announced in 1977 but he overrode his inspector and refused to allow the planning authorities more than token powers to control pollution, striking out the conditions relating to air pollution abatement and monitoring as duplicating the powers of HMIP, and weakened others. Nevertheless, there have subsequently been few environmental problems from the works comparable with the pollution experienced in the 1960s and early 1970s which had so incensed local residents. In the early 1980s a new prill tower was constructed utilising control technology which significantly reduced air pollution while permitting output to be increased. The level of complaint, both directed to the air pollution control authorities and expressed at the meetings of a local liaison committee set up after the early difficulties, has been very low in recent years and hedgerows have recovered.

The planning officers recommended approval in this case, subject to air pollution control conditions that were largely acceptable to the developer. UKF may well have exaggerated the difficulties and expense of monitoring because it was reluctant to reveal the relevant information to the public. Corporate interests proved triumphant, as did HMIP in having planning conditions relating to air pollution control deleted from the permission by central government. That the officers were initially overruled by the elected representatives (who were conscious of the project benefits and of union support for it) was largely the result of the influence of local objectors. These objectors, for whom this expansion was the first meaningful opportunity to protest about the fertilizer plant, were instrumental in having the application called in and in making the land use planning agency initially oppose the development. It is ironic that planning controls over pollution would probably have been stronger had the county been able to grant permission subject to the conditions negotiated with the company.

THE ST HELENS SULPHURIC ACID WORKS

Leathers Chemical Company Ltd sought planning permission to construct a major sulphuric acid manufacturing plant in St Helens, Lancashire in 1968[9]. One of the reasons for proposing to locate the works in a mixed residential/industrial neighbourhood, adjacent to an existing works, was environmental. It was suggested that sulphides arising from the existing works could be used as feedstock, thus reducing odours from a notorious water course due to sulphide effluents and bringing about a net decrease in existing emissions of sulphur dioxide to the atmosphere. The application contained little information about likely pollution arising from the sulphuric acid plant itself, or about the effects of such pollution on the environment of the works.

Nevertheless, the local planning authority was conscious of potential problems and sought the advice of local private consultants, but not of its environmental health department or HMIP. The consultants were favourably disposed towards the development but raised the problems of pollution caused by non-routine incidents such as the bursting of a pipe or the leaking of a valve. They proposed that arrangements for monitoring sulphur dioxide be made, that quantitative limits should be placed on discharges of sulphur oxides to the atmosphere and that the plant should close down if such limits were exceeded for more than 30 minutes. The planning committee, which did not have the benefit of many public representations, decided to grant permission for the development largely as a result of the net reductions in pollution anticipated. They incorporated the consultants' recommendations in the conditions attached to the planning permission. The use of the sulphides as feedstock proved, in the event, not to be feasible. HMIP accepted a chimney height of 43m, provided the foundations were such as to support a stack of 61m.

The works, once in operation, frequently emitted substantial quantities of sulphur oxides, especially during start-ups and during break-downs. These led to elevated ground level concentrations of acidic pollutants, which damaged fabrics (including ladies' stockings) and motor vehicle paintwork, caused severe coughing and led to a number of people being taken to hospital. Not surprisingly, pollution from the works became a major public issue, despite the absence of representations at the outset.

The local authority logged numerous serious pollution incidents and served an enforcement notice on the company for exceeding the emission level stated in the planning condition. A well-attended public inquiry was held into the appeal by the company against enforcement in 1973. HMIP appeared on behalf of the company and eminent counsel and witnesses were retained by both parties. The appeal was successful because the planning conditions were held not to be worded rigorously enough. The inspector recommended raising the chimney height. A prosecution of the company by HMIP was, however, successful but the fine eventually imposed was only £25.

Further incidents took place and, partly as a result of very significant public pressure by a local action group, St Helens council voted in 1975, after considerable debate, to serve a discontinuance order on the company notwithstanding likely costs of several million pounds. This action group had won extensive publicity in the local press and even managed to elect a member to the council of the local authority on a 'Leathers Out' ticket. At the instigation of HMIP, meanwhile, the company applied for planning permission to extend its chimney height to 61m. This was eventually granted and the chimney was erected in 1975, when a general overhaul of pollution equipment was undertaken. A second public inquiry was held in 1976, into Leathers' appeal against discontinuance, at which HMIP again supported the company. The inquiry inspector recommended that the appeal be upheld because the works could now be expected to operate more satisfactorily as a result of recent modifications (especially the chimney extension). He roundly criticised HMIP for its slowness to recognise the desirability of this overhaul and extension. The Secretary of State concurred with his inspector's decision on the discontinuance order.

Pollution incidents have occurred since this ruling. Though environmental conditions improved in the vicinity of the works, pollution levels remained generally unsatisfactory in a mixed residential area and the initial decision to locate the works in such an environment – despite the ostensible initial pollution abatement reasons – has been widely regarded as misguided (Miller and Wood, 1983). In the early 1980s the ownership of Leathers Chemicals switched to the Hayes Group, which has given much greater priority to pollution control. The company has invited councillors to inspect the works, has replaced much of the offending process plant and has seen its efforts rewarded by a reduction in the number of complaints to negligible proportions and the resignation of the anti-company councillor. Unfortunately, a major release of sulphuric acid vapour occurred in July 1986, which required the attendance of the emergency services and police warnings to residents to stay indoors and some admissions of employees and passers-by to hospital. This led to a prosecution by the Factory Inspectorate and payment of a fine of several thousand pounds. Subsequent further refurbishment has, however, reduced pollution significantly.

This case illustrates the necessity to undertake full and meaningful consultations before granting a land use permit for a major air pollution source. It also illustrates the difficulties of securing effective enforcement once a source has begun to generate pollution problems. The nature of central government control over this enforcement process is shown by its reluctance to sanction discontinuance – and by the air pollution control agency's support for the company. The air pollution control agency was, as the planning inspector pointed out, dilatory in pressing for the extensive modifications required and appears to have taken the company's part during the whole dispute, notwithstanding the prosecution it eventually felt impelled to bring. The power of objectors to influence the land use planning agency is clearly demonstrated, since the election of a single-issue councillor and the agreement to pay very high compensation were both unusual. Finally, the attitude of the company's management to pollution control, which appears to have changed markedly with the change of ownership, has had a notable effect on environmental problems and led to belated improvements.

THE GLOSSOP MOLYBDENUM SMELTER

A molybdenum smelter has operated since the 1930s in Glossop, on the outskirts of the Peak District National Park; a town in a basin almost completely surrounded by hills[10]. Ferro-Alloys and Metals Ltd had gradually expanded its operations in a mixed residential/industrial area since commencing manufacture. In 1970 planning permission for the construction of a new smelting furnace was granted by the local planning authority, after some consideration of the consequential environmental pollution. An air pollution control condition was incorporated. This decision, subsequently described by the Local Government Ombudsman as a 'serious error of judgement', involved the continued use of the company's 52m chimney stack to disperse sulphur dioxide laden gases after passage through an electrostatic precipitator.

By 1974 it had become apparent that the plume from the chimney was causing intermittent smell, taste and throat irritation and local residents, who boasted many amenity societies, were complaining bitterly about pollution from the smelter. Although molybdenum emissions were being satisfactorily controlled by the electrostatic precipitator, so far as could be ascertained, complaints about the effects of this heavy metal, as well as about sulphur dioxide, were prolific. HMIP, to whom many complaints were addressed, suggested that a much taller chimney (90–120m) was the only answer and the firm duly applied for planning permission for a 120m stack.

Despite contrary advice from HMIP, the new post-local government reorganisation planning authority (High Peak Borough Council) did not feel that sulphur dioxide concentrations were sufficiently prejudicial to health to justify the visual intrusion of such a large chimney in an attractive town and refused permission in 1975. The environmental health department concurred with this decision as there had been only a limited number of monitoring readings at the time. However, the number of complaints continued unabated, HMIP persisted in pressing for a taller chimney and the environmental health department began to gather detailed evidence of elevated sulphur dioxide levels.

A residents' petition was received by the council and a number of heated public meetings took place at which the public expressed opposition to the continued operation of the company because of the pollution it was causing. The dangers of heavy metal pollution were stressed again and again. Accordingly, after an unpublished investigation of alternative means of control of sulphur dioxide by the environmental health department had found no practicable means of removal of sulphurous gases, the chief planning officer invited the company to resubmit its planning application for the tall chimney in 1976.

The company promptly submitted an application for a 90m chimney (with scope for extension to 120m if necessary) which met with concerted opposition from the local residents who believed that alternative means of sulphur dioxide removal existed, that sulphur dioxide concentrations would continue to be a problem despite the gas being dispersed from a greater height and that the visual impact of the chimney was unwarranted. Another petition was presented but, in 1977, the planning committee granted a temporary 10 year planning permission after considerable discussion. In view of the continuing disquiet expressed by the unprecedentedly active residents – who managed to obtain national publicity for their campaign against the company – meetings with pollution control equipment manufacturers, and with the Secretary of State for the Environment, were held without altering the decision.

The residents served a High Court writ on the company to reduce pollution but this was later dropped because of the costs involved. As mentioned earlier, the Local Government Ombudsman investigated the whole affair but found no maladministration by the planning authority though he questioned the merit of the decisions taken. The chimney was built and improved the situation, although residents further afield were then subjected to occasional groundings of the plume on the hill slopes. The possibility of raising the chimney height further was discussed but not implemented. The number of complaints dropped but, as the 10 year temporary permission period drew to a close, increased again.

An intensive sulphur dioxide monitoring programme was undertaken by the local environmental health department in preparation for the reapplication which demonstrated that, although daily readings no longer exceeded the European Commission standards, major short-term episodes did occur. Following involvement by the Chief Inspector, HMIP served an improvement notice on the company giving it five years to reduce emissions. When the company reapplied a public meeting was held which, unprecedentedly, attracted nearly 1000 objectors. As a result of this expression of opinion, the full council refused permission for the continued use of the chimney and High Peak Borough Council served an enforcement notice requiring its removal.

Ferro-Alloys appealed against both decisions and a public inquiry was scheduled for late 1988 to determine the fate of the works.

This case illustrates the difficulties faced by a land use planning agency once permission to locate a source of air pollution has been granted. This was a classic British case in which a higher chimney was deemed to be the only method of reducing pollution which was technically feasible. Removal of sulphur dioxide by scrubbing was initially ruled to be impracticable by HMIP, despite evidence to the contrary presented by objectors. The objectors may well have been trying to raise the developer's control costs, to force him to relocate. The conflict between visual amenity and air pollution control was nicely exemplified by the land use planning agency's initial refusal to permit a higher chimney, despite HMIP's advocacy of it. The developer was able to exploit divisions between these two agencies to postpone taking action. It was the emotive campaign of the objectors which eventually forced the land use planning agency to accept a higher chimney reluctantly and temporarily. This has certainly led to a reduction of ambient concentrations, though not the removal of problems. The detailed investigation of these, and the continuing campaign by the objectors (which attracted many more supporters), were instrumental in the council's refusal to extend the life of the planning permission.

NOTES

1 Other British cases illustrating the use of both air pollution and land use control powers have been written up in detail: Holme Pierrepoint power station (Gregory, 1971); the Anglesey aluminium smelter (West and Foot, 1975); The Manchester steel mill (Wood and Pendleton, 1979); and the Bedfordshire brickworks (Blowers, 1984).
2 Central Directorate on Environmental Pollution (1982: 16). The Government has, in the past, conceded that 'in exceptional cases' planning conditions restricting, for example, the sulphur content of fuel oils consumed in registered works and the hours of operation of scheduled processes might be justified (Department of the Environment, 1972).
3 Anonymity was promised to the company and the new town development corporation in undertaking the case history and is preserved in this summary, which is partly derived from Ledger (1982).
4 This account is derived from Ledger (1982).
5 Under this act districts could refuse, but not approve, a county matter. The case history is derived from Ledger (1982).
6 This account is partly derived from Miller and Wood (1983).
7 This account is partly derived from Miller and Wood (1983).
8 This account is partly derived from Miller and Wood (1983).
9 This account is partly derived from Miller and Wood (1983).
10 This account is partly derived from Ledger (1982).

7
Outcome of the siting process for an air pollution source

This chapter contains an analysis of the roles of the various actors in determining the outcomes of the siting process for new stationary sources of air pollution exemplified in the case studies. Examinations of the authorisation process as a whole, and of the implementation of controls are also presented. The discussion of this chapter deliberately focuses on the common elements of the approval processes in America and Britain during which anticipatory air pollution controls are utilised. There are, of course, very substantial differences between the United States and the United Kingdom which are explored in detail in Chapter 8.

There have been several relevant previous studies of the siting of controversial developments, of which Gladwin's (1980) study of 366 site-specific environmental battles in the United States is perhaps the most instructive. He reported a change in focus from existing to greenfield projects as targets of environmental concern:

> environmental conflict is focusing on change – the focus is shifting from old to new targets, from existing pollution problems to potential environmental impacts and from 'band-aid' remedies to preventive or risk reduction measures.

He concluded that the notion that there is no best way to manage environmental disputes has begun to gain wide acceptance. Gladwin also believed that the outcome of an environmental conflict could not yet be predicted on either a theoretical or an empirical basis. He indicated that dozens of interacting variables served to shape the magnitude and distribution of the conflict outcomes. The most important factors in the cases he examined were probably the:

> characteristics of the parties involved, nature and magnitude of the goals in contention, nature of the issues at stake, past and anticipated relationship between the parties, strategies and modes of conflict behaviour engaged in, differential power or resources among the parties, presence and influence of audiences, availability and use of third parties, and character of the resolution mechanisms employed.

Seley (1983) has also investigated siting decisions and identified 150 relevant dimensions determining their outcomes (pp 26–31). Unfortunately, these were too numerous to be helpful in predicting the outcome of the siting process. He concluded that:

> A single community leader or a particularly insensitive agency official may mean more to the eruption and outcome of a controversy than all the theories of political and social behavior. The continued inability to predict or forestall conflict speaks to the basic truth of this fact (p 196).

Some of the conceptual frameworks advanced to explain the policy implementation process[1] could, perhaps, be applied to the siting of new air pollution sources. Mazmanian and Sabatier (1983) in their examination of policy implementation, suggested the types of variable likely to be significant in any particular instance: the tractability of the problem; statutory and procedural influences; and personal influences on the decision. Their framework stressed the need for clear and mutually consistent objectives in the statutory policy to be implemented[2]. Unfortunately, if all the variables necessary to account for a particular decision on whether or not to approve a new air pollution source are set down, the model they suggested becomes hopelessly unwieldy for explanatory or predictive purposes.

Further, the requirement for clearly defined policy objectives is hardly ever met in practice, as Fudge and Barrett (1981) have pointed out. They, too, tended to emphasise the role of negotiation in achieving objectives and asserted that:

> observation of what actually happens in practice leads to the inescapable conclusion that certain individuals, groups or governments tend to find a way of doing or getting done, what *they really want to do* while others do not (p 275, emphasis in original).

Majone and Wildavsky (1979) drew rather similar conclusions. They stated that 'implementation is evolution'. In practice, implementing a policy is a unitary process or procedure, not a tandem operation of setting a goal and then enforcing the plan that embodies it (p 180). They could have been writing about the nature of policies relating to new source control embodied in the air pollution legislation when they concluded that:

> In most policies of interest, objectives are characteristically multiple (because we want many things, not just one) conflicting (because we want different things), and vague (because that is how we can agree to proceed without having to agree also on exactly what to do). So if the objectives are not uniquely determined, neither are the modes of implementation for them (pp 182–183).

If this is true of the air pollution control legislation, it must be much more true of the land use control system, where objectives are far more numerous and decisions less clear-cut.

142 *Planning Pollution Prevention*

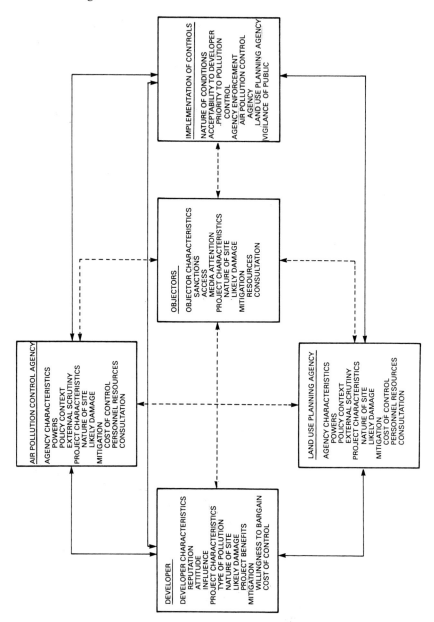

Figure 7 The main factors involved in determining the outcome of the new source siting process

Bardach (1977) has argued that the implementation of a policy is prone to the diversion of resources, the deflection of goals, dilemmas of administration and the dissipation of energies, which he termed 'implementation games'. There are frequently delays in the 'game'. He stated that:

> The implementation process is (1) a process of assembling the elements required to produce a particular programatic outcome, and (2) the playing out of a number of loosely interrelated games whereby these elements are withheld from or delivered to the program assembly process on particular terms (pp 57–58).

This analysis, stressing the control various individuals hold over the elements in the implementation process, helps to explain why negotiation appears to be such an important element in the application of conditions to new stationary sources of air pollution to minimise damage.

It would thus appear from the work of Gladwin, Seley, Mazmanian and Sabatier, Fudge and Barrett, Majone and Wildavsky, Bardach and others[3], that no simple formula is readily available for predicting the outcome of any case history involving anticipatory controls over air pollution. The most that can be attempted is the postulation of a number of indicators of the likely outcome of a particular siting decision. This may be done by examining the roles and characteristics of the developer, the air pollution control agency, the land use planning agency and the objectors in the siting process, then considering the siting process as a whole and finally analysing the implementation of controls on those new sources which developers are permitted to construct. Figure 7, which follows the outline of the siting process illustrated in Figure 1, summarises many of the factors influencing the outcome of the process.

THE DEVELOPER

The degree of conflict arising between a developer and the control agencies or sections of the public will depend, among other parameters, on the developer's attitude and responsiveness and on the general local climate of opinion. Few developers will seek conflict. As Gregory (1971) puts it: 'few leading businessmen or members of public boards are totally unmindful of society's informal sanctions and constraints' (p 297). Royston (1979) makes much the same point about the benefits of industrial concern for the environment:

> Environmental conflict... is sparked off primarily by fears, justified or otherwise, about the pollution which will be caused by a given development. Conflict can be expensive, and it can and does destroy projects and companies. If survival is one of the prime objectives of the enterprise then it must avoid conflict of all kinds, with local communities, with organised groups, and with government – at local, regional or national level (p 56).

Developer characteristics

The reception a prospective developer will receive from the local community will depend partly upon his or her reputation, attitude and influence. If the developer's enterprise is a local company or organisation, with a reasonable environmental record, or perhaps a company with a good environmental *reputation* nationally, he would be expected to be received more favourably than an unknown organisation from another part of the country or a company with a poor record. Chloride's local reputation in Bolton was undoubtedly one reason why opposition to its new lead battery plant was muted. On the other hand, the suspicion with which Palmer Barge Line from Texas was regarded in neighbouring Louisiana when it sought to build a creosote storage plant represents a good example of parochial antagonistic attitudes.

The developer's reputation will be particularly important if he is contemplating the expansion of an existing source which has given rise to pollution problems. Thus, proposals to expand Spectron's Providence solvent recycling plant in Maryland were poorly received as a result of the pollution previously caused by the company. Similarly, the poor environmental reputation of Sterling Mouldings in Tameside was undoubtedly the reason for strong local opposition to its proposed resin manufacturing plant.

The *attitude* of the developer, particularly in the early stages of the permitting process, will be crucial. If he can show himself to be accommodating in seeking to meet objectors' fears or valid criticisms by modifying his proposal, if he can provide sufficient information and appear open to all criticisms and suggestions, he is likely to be more successful than if he appears intransigent and to regard his project as an unqualified blessing for the chosen community (Storper et al, 1981; O'Hare et al, 1983). The Petroplex asphalt plant in Texas furnishes an example of a company refusing to concede an inch to potential opposition and paying for its intransigence with prolonged conflict. The Chloride lead battery works in Bolton, on the other hand, demonstrates what can be achieved by a more co-operative and open approach to comment and suggestion.

As Blowers (1984: 239–240) has stated, corporate interests have a large measure of effective power, or *influence* which they can wield by the presentation of evidence, by exploiting divisions and making concessions, by building alliances with other interests and by threatening investment withdrawal. Examples of all these tactics can be quoted from the case studies. Chevron's ability to present credible evidence was apparent throughout the California oil refinery application. Ferro-Alloy's exploitation of divisions between the Glossop Borough Council and HMIP was notable in the molybdenum smelter case. The new town glass fibre works case study provides an example of a developer making several concessions during the approval process. Dow

Chemical's forging of common cause with Solano County in California over the building of the chemical plant is a typical example of alliance building. Finally, UKF used the tacit threat of withdrawal from its Cheshire fertiliser plant in its negotiations with the planning authorities.

Project characteristics

There are several specific local factors which can militate for or against the success of a development proposal: the nature of the pollutant; the nature of the site; and the likely damage. If the *type of pollution* is particularly emotive then it will reinforce local residents' fears and mobilise opposition. There are certain 'political' pollutants: polychlorinated biphenyls, dioxin, asbestos and so forth which 'will automatically raise the visibility of the debate due to widespread public recognition of their harm' (Morell and Magorian, 1982: 124). Industry's assurances tend to be discounted where these pollutants are involved, due to well publicised previous incidents. There will probably be little 'hard' scientific evidence firmly to prove a case, and any decision in these circumstances may carry an element of risk. Examples of this type of pollutant are dioxin in the Oregon energy recovery facility case and creosote in the Louisiana storage plant case (where creosote was locally controversial).

The *nature of the site* will be a determinant of the response to the developer's proposal. If this is in an existing industrial area, or if an extension or modification is proposed on an existing firm's land, opposition will usually be muted (unless the area or firm have poor reputations). Applications for modifications or extensions to existing plants (especially large plants) are thus usually approved (for example, Chevron's oil refinery modification in California, and Ferro-Alloys' molybdenum smelter in Glossop were granted permits altering existing plants despite having unconvincing reputations). Only Spectron, in Maryland, and Sterling in Tameside, were refused permission to expand their plant (temporarily), largely as a result of their previous records. Development on green field, or previously non-industrial, sites is likely to be more contentious. It is notable that the Texas asphalt plant, the Bolton incinerator, the Louisiana creosote storage plant, the Oregon energy recovery plant and the California chemical plant, all on green field sites, were refused the necessary permits.

The level of *likely* pollution *damage* will depend partly on the nature of the local terrain and the proximity of homes, schools, hospitals, etc. While it is very difficult to calculate this level of damage, judgements can be made. Local opinion may exaggerate it, just as the developer may under-emphasise it. Thus the pollution from the Dade County resources recovery plant in Florida was not expected to cause great damage because of its distance from sensitive receptors and consequently it attracted no initial opposition. On the other hand, the problems caused by Ferro-Alloys' molybdenum smelter in Glossop have been exacerbated by the location of the town in a bowl surrounded by hills and led to marked antagonism when a planning permission was sought.

The developer will usually have to convince both the control agencies and local residents of the need for the project as well as of its environmental acceptability. A major new source of air pollution normally brings *project benefits* as well as environmental and social costs to the locality chosen. The developer will frequently be able to offer a substantial amount of additional employment in both the construction and operating phases of the project's life and will certainly make a contribution to the tax base of the area concerned. He can therefore often expect encouragement from the local authority. The attractions of the BECO oil refinery, North Carolina, of the Oregon energy recovery facility and of the Dow chemical plant in California to the small local authorities concerned were not unconnected with their anticipated local tax yields. Labour leaders, local businessmen, and other such economic gainers may also lend support. This was evident, for example, from the Citizens for Common Sense group in Portland, Oregon for the energy recovery facility and from the Transport and General Workers' Union for the fertilizer plant extension in Cheshire.

It is noticeable that, in those cases where serious disputes arose between the land use planning agencies and the developer, the benefits of the project for the local community were relatively small. For example, the Palmer Barge Line creosote storage plant in Louisiana, and the Petroplex asphalt plant in Texas, offered only relatively small advantages. Conversely, the California oil refinery, notwithstanding its poor environmental record, was too important to the City of Richmond to merit any challenge.

Mitigation

The importance of mitigation measures in resolving environmental conflicts has been stressed again and again, for example, by Duerkson (1983). On the one hand, the developer will obviously wish to build the new facility as inexpensively as possible. On the one hand, the developer will obviously wish to build the new facility as inexpensively as possible. On the other, he will be aware that the development will give rise to various environmental impacts and that he will be expected to spend money to mitigate these. Thus, most developers will expect to meet all the legally specified environmental control requirements relating to a proposed project to gain their permits.

A developer may, however, decide that it is in his interest to mitigate the effects of a new plant to the point where conflict is avoided. This will frequently mean both incorporating environmental controls to meet legal requirements in the first instance and making further concessions on environmental and other mitigation measures later on. This was the strategy successfully employed by, for example, BECO in its oil refinery application in North Carolina and by UKF in its fertilizer plant extension in Cheshire.

Thus, the developer may incur voluntary expenditure to make the new facility more acceptable. This may be achieved, for example, by visual amelioration or by paying for certain off-site improvements not required by law. An example of this would be Chevron's payment of $750,000 towards traffic management measures and road improvement design studies in the course of obtaining its oil refinery modification conditional use permit from the City of Richmond, California. Acceptability may also be gained by reducing air pollution emissions below the statutory levels. Chloride's voluntary proposal of lead emissions well below HMIP's presumptive limit in constructing a new battery works in Bolton exemplifies this.

While it will be perfectly obvious that the developer has to be *willing to bargain* with the air pollution control, land use planning and other agencies, it may be much less clear that he may also have to negotiate with various local groups. New voices of opposition may be heard when the developer feels the situation is under control (O'Hare et al, 1983). Examples of this phenomenon are the springing up of groups opposed to the proposed Spectron solvent recycling plant at Childs, Maryland, and of the election of a councillor in St Helens solely on the strength of opposition to the existing Leathers sulphuric acid plant.

The developer will thus often try to identify opposition leaders and, where possible, seek to negotiate with them. Negotiations can be undertaken directly or, more usually, through intermediaries such as the land use planning agency. For example, the North Carolina Department of Natural Resources and Community Development attempted to set up a committee, involving the expected opponents of the BECO oil refinery, to facilitate direct discussion of mitigation measures.

There are, however, financial limits to mitigation. The *cost of control* may become too large. If the proposed level of control will still result in what are perceived to be very high emissions, resistance to the proposal will remain severe. This was certainly true in the molybdenum smelter case in Glossop where Ferro-Alloys had claimed that no reduction in sulphur dioxide emissions was technically feasible. However, only a few years later in Portland, Oregon, Metro was able to offer 80% removal of sulphur dioxide from the energy recovery facility emissions by the use of a scrubber[4]. The concept of feasibility generally appears to include financial as well as technical considerations.

To eliminate all pollution would, in most cases, render a plant uneconomic. Generally, both statutory and additional control costs follow the same pattern. Relatively large amelioration for relatively low expenditure is usually followed by rapidly increasing costs for relatively small degrees of further control. Thus even BECO, which set out to be as co-operative as possible in gaining permission to build its oil refinery in North Carolina, drew the line at the expense of the state's condition relating to epidemiological studies.

There comes a point when the total cost of mitigation becomes too great and threatens a project's viability. Thus, while the precise figures may not be known publicly, the developer is normally only too aware of the financial constraints on the mitigation measures he can offer. As Morell and Magorian (1982) stated: 'the developer can simply abandon his siting proposal if, for example, local demands for compensation are too high or state regulatory requirements are too onerous' (p 124). The Dow withdrawal of its applications to construct a chemical plant in Solano County, California furnishes a rare example of such an abandonment, ostensibly because of onerous regulatory requirements.

A further financial cost to the developer to be taken into account is that involved in the permitting process itself – the transaction costs. If delays occur in the expected authorisation process timetable, these will result in effects on corporate strategy, in increased uncertainty in financing, in increased construction costs and in changing market conditions (Morell and Singer, 1980). Thus, Metro spent over $2M on the protracted permit application process in Portland, Oregon, before it abandoned its proposed energy recovery facility. The longer the delay, the more expensive it is likely to become, though delays may sometimes reveal that a project was misconceived, as, for example, in the Palmer Barge Line case where the plant was never constructed in Mississippi because of the downturn in the economy.

All these factors influencing the likely outcome of a developer's application to build a new stationary source of air pollution can be summarised in an indicator which appears to be true in the case studies analysed:

> The more open the attitude of the developer, the greater the developer's local influence, the greater the benefits offered by the project, the more the developer is prepared to make meaningful concessions to the control agencies and local residents, the more industrial the previous use of the site and the surrounding area, the further the site from existing homes and community facilities and the less 'political' the pollutants, the better is the chance of the new source being constructed and operated.

THE AIR POLLUTION CONTROL AGENCY

The role of the air pollution control agency in the authorisation process for a new stationary source of air pollution is to determine whether or not to grant a permit, and to impose anticipatory controls by means of conditions to any permit, so that the new source meets the legal and administrative requirements in force. This will involve imposing conditions on emissions by using equipment performance standards, emission standards, air quality standards, and other means (Chapter 2). Another function of the agency is to ensure that newly constructed sources (and other existing sources) meet whatever conditions are imposed, or general legal requirements, by means of various enforcement measures.

Agency characteristics

Air pollution control agencies are basically single purpose organisations. Although their general goal is to reduce air pollution or to prevent it from increasing significantly, the legal *powers* at their disposal are normally specific, though they vary. The legal and administrative requirements may be national, regional or local, but they will normally leave the air pollution control agency scope to negotiate more stringent pollution controls than the minima specified.

The air pollution control officials designated to deal with the permitting of new sources are normally engineers and, as O'Hare et al (1983) put it:

> The basic philosophy of the profession is that if two engineers disagree about the best way to do something, a few hours with a computer and a blackboard will result in their agreement on a single approach (p 27).

It follows that the air pollution controllers will generally be able to agree a set of conditions for a new source provided the developer is prepared to expand the necessary resources, or adopt appropriate practices, to ensure that the rules are met.

Provided a new source meets certain criteria (eg, emission levels or performance standards) there will be no grounds for refusing it the necessary largely non-discretionary permits. Of all the projects studied, only Dow Chemicals' chemical manufacturing plant in California was refused an air pollution permit and then only because no offset rule was in operation at the time. Even in the creosote storage plant case in Louisiana, where exemption was first granted and then denied, there was never any doubt that the air pollution permit would have been granted in the end. In the new town glass fibre works case, a permit was eventually granted after agreement between the controllers and the developer, despite the expression of concern by HMIP.

Even if a source is refused in the first instance, providing the developer is prepared to bargain seriously, the air pollution control agency will eventually grant the necessary air pollution discharge permit if its rules permit, though there may be considerable delays and compliance may be costly (Morell and Singer, 1980:193). The seeming rigidity of the US ambient air quality standards may eventually mean that air pollution control agency rules will become less flexible. However, experience suggests that inflexibility in air pollution control, once demonstrated, is soon successfully challenged. The introduction of offset arrangements as a result of the Dow Chemicals manufacturing plant case in California is an example of this process of regulatory amendment and accommodation.

Although most air pollution control agencies are unitary, they are usually subject to at least an element of political control. The way in which the relevant rules are interpreted will determine the *policy context* of the agencies. Thus, the British environmental health authorities are answerable to a committee of elected members and most American pollution control agencies are answerable to boards of nominated members. This political control is frequently somewhat remote from public accountability (the British central Inspectorate is especially independent of political and thus public control) and hence not as easily susceptible to pressure as the land use control agency.

This element of political control, however, ensures that objectives other than reducing air pollution become important in the work of the agencies. For example, political representatives may be well aware of the necessity not to lose existing sources of employment and revenue, and to attract new ones. The financial costs to the developer and the social costs to the community of controlling pollution will thus often become very explicit in the agency's decision making. For example, the appointed Secretary of the North Carolina Department of Natural Resources and Community Development stated in public that all the necessary air pollution and other permits for which his agency was responsible would be forthcoming in the interests of local economic development.

The effectiveness of *external scrutiny* in ensuring appropriate implementation of legal and administrative requirements by the agency will be dependent on the formal public participation arrangements and on the amount of information available about its activities. Thus, several American air pollution control agencies increased the stringency of their conditions as a result of public pressure. Examples include the Bay Area Air Quality Management District's controls over the Chevron California refinery modification. Similarly, the politically responsive nature of the Bolton environmental health department helped to achieve stringent controls at Chloride's new lead battery works.

The degree of such external inspection is a crucial determinant of the nature of the relationship between the air pollution control agency and the developer. If a decision is likely to be scrutinised by the courts or at appeal, the temptation may be to play it by the book and perhaps to eschew innovation or experimentation. This was the situation in the North Carolina oil refinery permit application case where the air pollution controllers were anxious to be seen to be following the letter of their rules scrupulously because of its high public and media profile.

This scrutiny can serve to prevent an agency developing too sympathetic an attitude towards the industrialist's position. A close relationship between the controller and the controlled, without a strong external moderating influence, can lead to decisions which favour the polluter rather than the potentially polluted. This was probably the case where HMIP, using the technically unconstrained and procedurally confidential best practicable means approach, supported Leathers Chemicals in the St Helens sulphuric acid works case and Ferro-Alloys in the Glossop molybdenum smelter case when both companies' operations were giving rise to serious pollution problems.

Project characteristics

While suitable anticipatory technical controls may be agreed, the unitary nature of air pollution control agencies ensures that they will not usually consider the *nature of the site* of a new source (proximity to receptors, local topography, etc) in setting conditions, even if they may be aware that local circumstances can be very important in determining *likely damage*. One reason for this is that the rules implemented by air pollution control agencies are not usually subtle enough to take location into account. Indeed, reference to land use considerations was deleted from the US Clean Air Act in 1977 (Chapter 3). The approach of the director of the Bay Area Air Quality Management District epitomises this eschewal of locational considerations: 'We don't care where a plant goes as long as it meets the standards. You could put a coal-fired plant in downtown San Francisco if it can meet the standards'[5]. In the United States the levels of air pollution in an air quality region (which may extend for hundreds of miles) determine the rules to be applied while more local considerations are largely ignored. The sole concession to locational considerations which is generally made in the United Kingdom by air pollution controllers is in regard to chimney height modification.

A second reason for ignoring the location of sources is that agencies would regard it as inequitable to penalise the developer of one new source by imposing stricter conditions than upon another with a similar plant, simply because of his choice of site. The air pollution control agency alone, therefore, cannot implement a policy of minimising pollution damage. It will, in practice, leave any decisions as to the appropriateness of a location to others (usually the land use planning agency). This is exemplified by the creosote storage plant case in Louisiana where the air quality section of the Department of Natural Resources postponed consideration of the grant of its air pollution permit until the coastal management section had completed its deliberations on land use issues.

Mitigation

The prevalence of negotiation in the implementation of air pollution controls has been stressed by several authors. Thus Downing (1982) stated, in considering various national systems that:

> Bargaining between the source of pollution and the responsible control authority is universal.... It is most often informal, typically technically illegal, and done in secrecy.... The result of this complex bargaining depends upon the relative strengths of the participants and the cost to each of participating in the bargain.

According to Downing, laws which ignore costs will be compromised and delay, consideration of technical issues and economic impacts, voluntary compliance and reluctance to resort to penalties are found in all cases.

The importance of bargaining in the American air pollution control system has also been emphasised by Hagevik (1970), who advanced a bargaining framework on the basis of American research. Indeed, he stated that 'most of the legislation is basically similar to zoning legislation' (p 6) where negotiation was also the accepted means of procedure. Certainly, negotiation took place in all the case studies where major emission sources were involved.

It is possible to visualise these negotiations with the polluter as the air pollution control agency trying to increase the level of air pollution control above the legal minimum requirements and the developer usually resisting this, because his capital and/or revenue *costs of control* would be correspondingly increased. As the parameters determining the point of compromise are so numerous, it follows that there will be considerable variations in performance between agencies, and perhaps between the outcomes of siting applications for similar new sources within the same agency's jurisdiction. Examples of such variations, admittedly over time, are seen in the particulate emission limit set by the Florida Department of Environmental Regulation on the resources recovery plant in 1977 (0.08 gr/ft^3); the EPA permit for the Oregon resources recovery plant in 1980 (0.04 gr/ft^3); and the Oregon Department of Environmental Quality's proposed limit of 0.015 gr/ft^3 in 1982 for the energy recovery facility.

The scope for negotiation by the air pollution control agency, and its success in controlling new sources of air pollution, will be affected by the attitude and quality of its *personnel resources* and by its procedures (including *consultation* with other bodies) as well as by the laws, regulations and guidance under which it operates. Thus the weakness of the staffing of the Louisiana Department of Natural Resources resulted in vacillation by the air quality division over the permit exemption in the creosote storage plant case. Similarly, the failure to consult adequately led the Tameside environmental health department to make unsubstantiated and erroneous judgements about the pollution to be expected from Sterling Moulding's new resin manufacturing process.

The outcome of the air pollution control agency's consideration of a developer's application to construct a new air pollution source can be expressed in the form of an indicator which appears to be true from the evidence of the case studies analysed:

> Negotiations between the air pollution control agency and the developer may be protracted but will almost always result in the granting of a permit with the precise degree of control exercised varying, and being more stringent where there is significant external scrutiny, but characteristically taking little or no account of the geographical site of the air pollution source and hence of the actual damage likely to be caused in that location.

THE LAND USE PLANNING AGENCY

The essentially local land use planning control agency has multiple objectives which will usually include the promotion of suitable employment sources and local revenues, minimisation of consequential public financial costs, the amelioration or preservation of the visual environment and the minimisation of air pollution (though they may not all be explicitly stated). All proposed new stationary sources of air pollution will involve the consideration of several sets of conflicting land use planning objectives. Further, it is unlikely that the existing policy of the agency will have taken account of the ramifications of the proposed source.

There are two distinct processes at work in the land use planning agency in considering a new air pollution source. The first is evaluation of the general advantages and disadvantages of each proposal. This provides the basis for the second: negotiation with the developer to reduce the environmental and social costs while increasing the community benefits of the development. Both officials and, frequently, politicians are involved in these processes.

Agency characteristics

The importance of the nature and range of legal *powers* available to the land use planning agency was underlined in several cases. In the new town glass fibre works case pollution control conditions, unusually, were appended to the lease for the site in the absence of other controls. The inability of Kirklees Metropolitan Borough Council to take action until Crewe Chemicals had formally changed the use of its chemical formulation plant site is another example of the importance of the nature of the legal powers available. The annexation powers of the City of Midland, Texas, also proved to be important in resolving pollution problems from the asphalt plant.

Blowers (1980) has castigated land use planning as being a short-term incremental process precisely because the *policy context* seldom allows difficult decisions, such as those concerning air pollution sources, to be taken on any but an ad hoc basis. He concluded that planning is a political rather than a technical activity and that it essentially consists of a series of negotiations:

> Planning ... mediates various interests, seeks to achieve consensus and attempts to co-ordinate and guide activities to avoid future conflicts.... Within the limitations prescribed by the necessity to ensure the maintenance of the prevailing pattern of social relationships, planners exert considerable influence and power (p 37).

Support for these assertions may be found in the case studies. A previously adopted land use planning policy context appears to have had little influence on the various decisions involving the air pollution sources studied, except in the Bolton sewage sludge incinerator case where the provisions of the informal local plan were instrumental in preventing development. (Several authorities reacted to proposed developments by formulating general policies once applications had been received: for example Midland City Council in Texas on asphalt plants.) Wherever a major employer was involved, the local planning authority sought to grant permission, reflecting the company's local economic influence. This approach was exemplified by Richmond's response to the Chevron Californian oil refinery modification and by Cheshire's attitude to the UKF fertilizer plant extension. Mediation was a role commonly adopted by planners in many of the case studies. For example, in both Bolton cases (the lead battery plant and the sewage sludge incinerator) the planning officers were essentially acting as mediators rather than as initiators in the pollution control decisions.

The role of land use planners in dealing with the siting of a new facility has been caricatured by O'Hare et al (1983) thus:

> The reason opposition exists is that different groups have different values and experiences. The way to deal with it is by increasing public participation: get everyone together in public hearings; be sure that everyone's view is heard by everyone else, and that the government agencies in charge know how their constituents feel about things (p 28).

As they pointed out, however, opposition may actually harden when more facts are revealed. The constant harping on a particular type of pollution, perhaps because of regulatory requirements, can elevate its importance well beyond that merited in the circumstances concerned. Examples of this phenomenon are the way in which the effects of dioxin were dwelt upon in the Oregon energy recovery facility case and those of creosote were exaggerated in the Louisiana storage plant case.

Healy and Rosenberg (1979) expressed the view that there will always be winners and losers in the land use planning process:

> Unfortunately, even the most astute and sensitive planning will not be able to make everyone at least as well off as before. Some will gain and others lose, whether by unchecked development or by carefully conceived land policies (p 242).

Land use planning agencies, being predominantly local, are usually only too conscious of the need to take account of the local opinions and the political muscle of the potential losers from a new air pollution source. In both the USA and the UK, public access to the essentially discretionary land use control process is far greater than to the air pollution control process. This *external scrutiny* means that they are much more susceptible to public and media pressure than air pollution agencies.

The attitude of elected representatives is thus an important factor in determining decisions. This political responsiveness in environmental matters is largely independent of party politics in Britain, as in the USA. This responsiveness was evident in, for example, the reaction of the elected representatives in Midland, Texas, to the objections raised by the public to the asphalt plant. The response of the three Tameside councillors, who nearly persuaded the council to prohibit construction, in the resin manufacturing mill case provides another instance of political susceptibility to public pressure. In both instances the air pollution control agencies could be seen to be much less responsive to public pressure than the land use planning agencies.

Although the land use planning agency officials may often be minded to grant permission to an industrialist to construct a new stationary source of air pollution, there will be at least some uncertainty about the outcome because of the accessibility of local representatives to resident opinion. Thus, there are sometimes differences of opinion between professional planners and their political masters. The advice of the officers was overridden, for example, in the Bolton sewage sludge incinerator case and, at least initially, in the Cheshire fertilizer plant extension case. However, whereas a local government will frequently object at the behest of its citizens, it will often favour a development opposed by certain groups of residents.

Blowers (1982), in discussing an application to construct a new brick works in Bedfordshire, affirmed that opinion about potential pollution was frequently a crucial factor in the political siting process. He classified environmental issues as dormant, passive or active:

> Important environmental issues become active when conflict between powerful interests can no longer be suppressed or contained. The importance of issues is not necessarily intrinsic but politically determined.

In general, the outcome is less certain than in the case of the air pollution permit, especially if the site concerned is not allocated for industry in local plans because the land use planning agency's powers are more discretionary. Thus, in the cases examined, four of the 16 developments were refused land use planning permits (Spectron at Childs, Maryland; Petroplex at Midland, Texas; Palmer Barge Line in St Tammany, Louisiana; and the North West Water Authority in Bolton). In each case (except Childs, Maryland, where other factors supervened) the site concerned was not zoned for industry.

Though there may be appeals against the local land use planning agency's decisions and though other agencies may be involved, there is no doubt that the ultimate decision as to whether to permit a new stationary source of air pollution to be constructed rests with the land use control agency and not with the air pollution control agency. As Morell and Singer (1980) put it, in discussing the siting of new energy facilities in the USA[6]:

> In general, acquisition of local siting approval has been the most difficult hurdle. Local governments tend to act fairly rapidly on permit requests ... but they may decide to reject the facility (p 300).

Discretionary land use approvals have often not been forthcoming in the USA 'because local authorities would not balance local desires versus statewide interest in both environmental quality and industrial management' (Morell and Magorian, 1982:91).

Project characteristics

Because any evaluation is likely to prove inconclusive, notwithstanding the fact that the developer will often be an outsider and the objectors local, the prevailing ethos of land use planning will usually be to permit the development (Miller and Wood, 1983) but to negotiate to mitigate its impacts. This will not be the situation, however, where the proposed new source is obviously entirely environmentally unsatisfactory because of the *nature of the site* or palpably politically unacceptable because of the emotive nature of the pollutant involved.

The staff and elected representatives of land use planning agencies, having very little knowledge of pollution, and little inclination to acquire it, may too readily take the air pollution control agency's permit at face value. This may lead to neglect of the fact that consideration of the location of the source rests solely with the land use planning agency, as does the use of land use planning techniques for pollution control. Consideration of land use controls over a source may thus fall between two stools. This was, effectively, what happened in the St Helens sulphuric acid works case, where location close to housing areas led to avoidable pollution problems. Similarly, the opponents of the Childs solvent recycling plant in Maryland had to remind the local land use planning agency that the state air management administration's rules did not allow that administration to consider location in deciding to grant its permit.

Tensions may arise between the local land use planning agency and the air pollution control agency, particularly when the planners seek to refuse the development, or to impose more stringent conditions than the air pollution controllers because they wish to eliminate or reduce the *likely damage*. Jones (1975:211) has stressed also that conflicts often exist between levels of government in the control of air pollution due to the clash of wider and local interests and to personal differences. These were evident, for example, in the UKF fertilizer plant extension case and the Bolton lead battery works case. Here the land use planning authorities, with the help of environmental health officers, sought to impose planning controls which were fiercely resisted by HMIP because they usurped its powers.

Tensions may also arise, of course, between the land use planning agency and the objectors. Objectors may be seeking denial of the permit when the agency is minded to grant it, or may not agree with the control techniques used by land use planning agencies in seeking to reduce air pollution. Such tensions were evident, for example, in the objections to the grant of the land use permit in the North Carolina oil refinery case where land use conditions relating to air pollution were not applied and in the objections to the higher chimney suggested in the Glossop molybdenum smelter case by the land use planners after consultation with the air pollution control agencies.

Mitigation

Acceptable compromises between pollution and amenity are bound to be sought where issues are not clear-cut. This need to compromise has led to the universal use of bargaining and negotiation in land use planning practice. Thus, a set of conditions applied to a land use permit is often a record of the negotiations between the developer and the land use planning agency, together with the controls necessary to implement these. Certainly the concept of planning gain is no more than an agreement that the developer will contribute a number of additional amelioration costs in return for the grant of his planning permit[7]. As Willis (1982) put it:

> Planning gains are defined as the achievement of a benefit to the community that was not part of the original application and, therefore, not normally commercially advantageous to the developer.

Since negotiation is an activity which normally presupposes the consent of the regulated and mutual trust with regard to the exchange of information, third parties – frequently those most affected by a decision – are often excluded from the process (Strauss, 1978; Fisher and Ury, 1983). For example, Jowell (1977) stated that:

> It is generally accepted that the effective negotiation of planning gain must be conducted among a small number in an atmosphere of relative secrecy. Little or no public participation is considered possible.

This exclusion may lead to frustration among objectors left out of the bargaining process concerning a new stationary source of air pollution.

The land use planning agency will clearly have some difficulty in negotiating on behalf of all the groups it represents, will need to allow at least some public participation and will be concerned with a large number of items, many of which – including the *cost of control* of air pollution – will be complex. It will require expertise to be available either in-house or through a network to assist in the process of negotiation. The land use planning agency may seek further air pollution controls, over and above those decreed by the air pollution control agency, as well as amelioration of social, economic or amenity impacts.

The importance of mitigation in the highly political land use planning decision making process can be seen in most of the cases where approval was granted. Perhaps the most obvious examples are the agreements signed by Chloride and Bolton in the lead battery case, and by Spectron and Cecil County, Maryland, in the solvent recycling case, and the agreed Oregon City planning conditions requiring the installation of a scrubber on the energy recovery facility. Some conditions demanded of the developer may concern matters beyond the confines of his proposed site, as in the case of the Chevron contribution to road design plans in Richmond, California.

Important factors in reaching a decision are the financial and *personnel resources* of the agency, together with the attitudes and knowledge of officials (Wood, 1979). The non-availability of in-house specialist staffing may not be a problem for land use planning agencies in dealing with pollution if they have the necessary financial resources. Thus both Cheshire and the new town development corporation hired consultants to help them evaluate the nature and effects of new plants proposed in their areas. The City of Richmond and Oregon City dealt with the problem of expertise in a different manner by arranging for applicants to hire consultants to provide technical appraisals in the oil refinery modification and energy recovery plant cases respectively.

While few planning officers in the cases studied had any marked attitude to, or extensive technical knowledge of, pollution, many of those in the English land use planning departments were able to draw upon the expertise of their colleagues in environmental health departments (for example, in the Glossop molybdenum smelter case) to help to mitigate impacts.

Environmental impact assessment of the damage likely to be caused by projects was employed in the Florida resources recovery facility case (under the power plant siting act provisions), in the North Carolina oil refinery case, in the California oil refinery modification case and in the California chemical manufacturing facility case. In each instance it appears to have highlighted some environmental problems and led to the formulation of controls to help to mitigate these. For example, the air pollution controls finally imposed in the California oil refinery modification case by the air management district owed much to the EIA report.

Despite the variety of problems encountered in the case studies, the range of land use planning techniques of air pollution control (Chapter 2) employed to mitigate these was limited. Where conditions were imposed by the land use planning agencies, these mostly related to emission levels or to monitoring. Thus, the Oregon City conditions on the energy recovery facility and Bolton's conditions in the agreement on the lead battery works both involved emissions limitations and monitoring. Many land use planning agencies omitted to impose any pollution control conditions. For example, the final permission in the UKF fertilizer plant extension case and the Solano County permit in the California chemical production facility case had no appended air pollution control conditions.

The only example of siting with respect to topography was the proposed relocation of the Maryland solvent recycling plant from its valley site to a more open situation. Pollution from both the Bolton sewage sludge incinerator and the Oregon energy recovery facility would have been exacerbated by their valley sites.

The developers in several of the case studies carefully sited industry with respect to sensitive receptors. For example, both the Bolton lead battery works and the Florida resources recovery plant were located well away from the nearest houses. The Texas asphalt plant is a notable example of a proposal that could easily have been located on land in the developer's ownership further from a residental area. The decision not to permit any emissions at all at the locations concerned was taken, at least implicitly, in the four cases where land use planning permits were refused.

The use of buffer zones within the site boundary to reduce the effects of pollution occurred in only two cases: the retention of existing woodland at the North Carolina oil refinery site; and the use of a planted mound at the Bolton lead battery works. Landscaping conditions were used in several cases, such as the Florida resources recovery plant, but these were not designed to reduce air pollution concentrations.

The design and arrangement of buildings to reduce pollution was evident in several instances. For example, the number of stacks was reduced during negotiations on the Bolton lead battery works. Conditions requiring high chimneys – to disperse pollution to reduce its effects – were employed in, for example, the new town glass fibre works case and the Glossop molybdenum smelter case.

Miller and Wood (1983) found that the most important parameters in the evaluation of local planning authority performance in the control of pollution were the costs of the decision, the influence of the public, the attitudes of elected representatives, the effectiveness of consultations and the availability of policy guidance. The case histories provide considerable support for this analysis. No large enterprises were refused land use permits in the cases examined and consequently the costs of decisions to the local communities were generally low: the refusals meted out did not entail the areas concerned foregoing large revenue and/or employment benefits. The influence of the public on the land use planning authority's decision was perhaps most apparent in the Louisiana creosote storage plant case and in the Bolton incinerator case in which permits were refused by elected representatives following recommendations from officers which were either neutral or supportive of the projects. The effectiveness of agency *consultation* was evident in the Bolton lead battery plant case where environmental health officers were instrumental in negotiating agreement to very low emissions.

Ledger (1982) has confirmed many of these findings. She concluded that, generally, 'the achievement of meaningful planning control of pollution was largely dependent upon consultation and co-operation with other pollution control agencies' and that:

> The most significant restriction on the realisation of a thorough system of pollution prevention and control by planning in practice is connected with the attitudes of planning officers and local planning authorities towards implementing and enforcing these controls (p v).

The likely outcome of the land use planning agency's deliberations can be expressed in the form of an indicator which appears to be true in the case studies analysed:

> The land use planning agency ultimately decides whether or not to allow a new source of air pollution to be constructed and, because its evaluation will frequently prove inconclusive (reflecting conflict between its objectives), it will usually seek to approve the source after negotiating increases in its benefits and reductions in its air pollution impacts but will not utilise many of the planning control techniques at its disposal.

THE OBJECTORS

Someone will always object to a new stationary source of air pollution if he believes it will cause damage to him or something he values. While the precise site, type of pollution and locality will condition responses, air pollution is an emotive topic and a convenient focus for those who may also, or even primarily, object to a new project for a variety of other reasons (eg, traffic, visual impact or property values). As Morell and Magorian (1982) put it:

> Those who oppose a proposed ... facility for such legitimate reasons, however, may well realise that they can mobilise much more support for their cause from others by raising the specter of fear (p 23).

Frieden (1979) found that environmental opposition in California was sometimes a successful ploy by affluent existing residents to protect their own amenities:

> The regulatory process in this case turns out to be highly political, with the priorities and directions set mainly by influential suburbanites. In contrast to earlier capture of regulatory agencies by industry, this time the captors are local growth opponents and public officials guarding their own turf against newcomers (p 177).

Gladwin (1980) was able to distinguish nine different types of opponent to proposed developments, ranging from local residents to foreign governments. He categorised these as governmental or non-governmental, local or non-local and found that most conflicts involved two or more sets of opponents who had a tendency to form coalitions. He found that while most environmental groups were only involved in certain types of conflict local residents 'appear more willing to oppose anything posing a perceived threat'[8]. Thus local objectors may ally themselves with the local branch of a national environmental group, may join a pre-existing local pressure group or set up their own organisation[9].

Of the case histories examined here, only the Florida resources recovery facility, the new town glass fibre works and the St Helens sulphuric acid works went through the process from application to construction without significant objections being raised. However, in each of these instances objectors soon protested against the operation of the new sources. Only the California chemical works and the California oil refinery modification attracted the attention of national pressure groups (the Sierra Club and Citizens for a Better Environment respectively). Even here the representation of these groups was more akin to the only regional group entering the lists in the cases examined: the Oregon Environmental Council in the energy recovery facility case. Many of the other case histories show the intervention of local chapters of national groups (such as the National Wildlife Federation in the Louisiana creosote storage plant case), of pre-existing local groups (such as the residents' association in the Bolton incinerator case) and of groups specially created to contest particular developments (such as Carolina Coastal Crossroads in the oil refinery case).

Objectors will frequently use all the *sanctions* at their disposal, including legal challenges and manoeuvres, to induce procedural delays to achieve their goals in air quality controversies (Caldwell et al, 1976). Resort to the courts raises the costs borne by the developer and imposes uncertainties on him. Legal action may be initiated by the objectors themselves or they may be able to persuade the air pollution control agency, or more usually, the land use planning agency to take this step. For example, the residents of Glossop attempted to force the closure of Ferro-Alloys in the molybdenum smelter case by taking out a High Court writ, but had to abandon the procedure for want of money. Objectors persuaded the City of Midland to take legal action against Petroplex in the Texas asphalt plant case.

The threat of procedural delay can be useful to objectors in negotiating mitigation of environmental impacts or other concessions. This mitigation by negotiation is perhaps more likely with large companies that are conscious of their environmental reputations. As Kimber and Richardson (1974) stated:

> The fact that modern corporations are increasingly anxious to present a good public image can only assist environmental groups in applying successful pressure against development proposals (p 222).

Thus, BECO was very anxious to negotiate mitigation measures with the North Carolina agencies and with objectors to the proposed oil refinery in order to foreclose the possibility of court cases and lengthy appeal procedures. Similarly, one reason why Chloride agreed to stringent emission controls in the Bolton lead battery plant case was to avoid a refusal of planning permission and hence the delay engendered by a public inquiry.

Objectors may also employ quasi-legal social tactics such as pickets, or the staging of organised demonstrations and protests at public meetings. Action to close roads near the Texas asphalt plant by Midland County, at the behest of residents, fell into this category. Morell and Magorian (1982:95–97) have given the example of a trench being dug across the access road to a plant by a local government. Union action may sometimes be utilised to frustrate a developer. Some activity by objectors may cross the boundary into the extra-legal category. The attempt to arrange the razing of Spectron's Providence solvent recycling works in North Carolina was a tragic example of this type of action.

Objectors need adequate information on which to base their case and appropriate public participation arrangements to give *access* to the decision-making process (Caldwell et al, 1976). They also require time to utilise the information before the crucial decisions are made by the control agencies. The availability of information in the United States as a consequence of the freedom of information legislation and, where they apply, requirements for environmental impact assessment, are significant advantages for objectors seeking to marshal their case. Thus, Citizens for a Better Environment never had any difficulty in obtaining the information the group needed to counter Chevron's California oil refinery proposals, though Citizens did complain about lack of time to assimilate some of these data. On the other hand, objectors in the Tameside resin manufacturing plant case had neither the information nor the time they required to present a cogent argument before the initial decision had been taken.

The time at which information is provided is often crucial in determining the objectors' response to it. If people receive information before they have relinquished their neutral positions and before they have developed a negative impression of the information supplier, they may be open-minded about it. However, once they have taken a position on either the proposed project or on the information supplier, additional information may become almost valueless (O'Hare et al, 1983).

It is precisely because, in both the United Kingdom and the United States, the land use planning process is concerned with change and is relatively open to the public that public participation in the process of bargaining to limit air pollution is much greater than in the air pollution control process. Thus, the public planning inquiry into the fertiliser plant extension in Cheshire provided the local residents with a rare opportunity to voice their discontent with the previous performance of UKF because HMIP had provided no opportunity for participation in its control activities. The same was true of the residents' participation in the land use planning agency hearings in the Maryland solvent recycling plant case, though limited access to the state air pollution control process had been available. The liaison established between the North Carolina air pollution control agency and the local land use planning agency by the Little Elk Creek Civic Association in this case was an example of making use of the public's right of access both to administrators and to information to gain more stringent controls.

Caldwell et al (1976) contended that it is the right of local and broader action groups to determine the future of their own environment and that governments should respond to such groups. They felt, however, that elected representatives are often unable to represent the views of the public. It is clear from the case studies that the better the formal and informal links objectors have with elected representatives, the more chance they have of success. Gaining the support of individuals or organisations with political 'muscle' is clearly invaluable.

Some politicians will not need to be influenced by objectors but rather will themselves galvanise opposition. For example, the councillor in the St Helens sulphuric acid plant case and the commissioner in the Oregon energy recovery facility case were instrumental both in focusing the efforts of objectors and in leading the objectors' case. In other instances the elected representatives were swayed by objections into opposing and eventually preventing the establishment of new sources – for example, in the Louisiana creosote storage plant and the Bolton sewage sludge incinerator cases.

The environmental movement has, on the whole, been skilful in winning *media attention* to assist in putting over its viewpoint on both national and local issues. As Frieden (1979) stated: 'Environmentalists' ... vision of environmental quality attracts popular support, and they have shown exceptional skill in presenting their views to the media and to the courts' (p 175). Reporters naturally enjoy investigating controversies and will frequently sympathise with the objectors against a large corporation. The media tend to be more concerned about environmental issues than might be expected from their ownership, providing objectors with substantial advantages in pressing their case (eg. Lowe and Goyder, 1982:74–80). The more media exist in a locality, the more this is likely to be true.

Many of the cases investigated in this study demonstrate that objecting groups can secure considerable advantage if they are able to enlist the support of the local media. There is no doubt, for example, that the article in the *Washington Post* on the Maryland solvent recovery plant had a marked effect by instigating a formal investigation. Where large existing plants were involved (eg, the Chevron refinery, the Cheshire fertiliser plant) it was notable that the local papers tended to be more balanced in their presentation of objectors' and proponents' viewpoints and devoted less space to the issues than in many other cases.

Project characteristics

The number and type of local objectors vary with the *nature of the site*. The population density of the area concerned and the socio-economic status of the residents of the locality will be important in determining both the importance attributed to the likely pollution and the weight attached to the benefits brought by the industry. This determination will depend upon the existing nature of the area and the employment, affluence and aspirations of the residents. Thus, the predominantly well-to-do residents of St Tammany Parish, Louisiana, were not greatly interested in attracting new industrial development and objected to the creosote storage plant.

The more local and issue-orientated the government of the area, the more elected representatives will be susceptible to political pressure (Hall, 1982:xxi). Equally, objectors will generally have less power to exert such influence the fewer they are in relation to the population of the local jurisdiction concerned. For some groups of local residents, the disadvantages of a proposed project may always outweigh its advantages. This characteristic is likely to be particularly marked where they live across the local government boundary and hence share none of the local revenue benefits of the development while still suffering from the pollution. This explains the objections of, for example, the Chester and Vale Royal district councils to the Cheshire fertiliser plant extension and of the residents of New Hanover County and of Wrightsville Beach to the North Carolina oil refinery.

For some, generally those opposed on principle to industrial development, the perceived *likely damage* from the air pollution source will be such that no mitigation of environmental and other impacts may ever be sufficient to gain acquiescence[10]. It is, for example, difficult to imagine that the chairman of Residents for Unpolluted Neighbourhoods in the Maryland solvent recycling plant case, or the Oregon City commissioner in the energy recovery facility case, would ever have been satisfied no matter what concessions the developers might have made.

Kimber and Richardson (1974) suggested that:

> Most development issues involve a fine balance of economic against environmental benefits and it is one of the tasks of amenity groups to present policy-makers with a balance of argument which is weighted less in favour of the economic interests than has hitherto been the case (p 215).

Theoretically, public decision-making demands that rational argument be advanced. However, there may well be tactical advantages to objectors in employing hyperbole and even misrepresentation about the likely damage, especially in the early days when opinions are formed. As Badaracco (1985) has stated: 'To exercise some form of indirect influence, a group must make itself heard.... Moderate, reasoned, tolerant statements are far less potent than adversarial sound and fury' (p 144).

An interesting feature of several case studies is that some of the most satisfactory outcomes for objectors arose when one group concentrated on rational argument while another resorted to hyperbole. The activities of the Little Elk Creek Civic Association and Residents for Unpolluted Neighbourhoods in the Maryland solvent recycling plant case and of the Oregon Environmental Council and the city commissioner's group in the Oregon energy recovery facility case can certainly be categorised in this way.

Mitigation

While the objective of many objectors will be to stop a project, their fall-back position will often encompass the possibility of additional mitigation. However, as O'Hare et al (1983) expressed it, local objectors frequently have:

> no reason to expect the developer to change his mind, alter his project, choose another site, or heed the public's concern. In fact, they perceive themselves as only having power to delay or stop the project ... (p 7).

They went on to assert that, while many developments were universally agreed to be necessary, objectors would oppose them by every method available because they were perceived to be *locally undesirable land uses* (LULUs). They would insist that they should be sited anywhere, but *not in my back yard* (NIMBY).

To achieve their aims, objectors require appropriate financial, technical, organisational and personnel *resources* and effective leadership. The involvement of existing environmental groups usually confers better access to these resources. Nevertheless, typically, objectors group together very rapidly and whether in existing, resuscitated or new organisations, are able to acquire sufficient information, knowledge, financial resources and skills to engage in meaningful, if sometimes unequal, battle against a developer. For example, the residents in the area around the proposed Bolton incinerator grouped together to pool their expertise and to raise financial resources.

Despite the ability of objectors to marshal sufficient resources to mount a campaign, their relative disadvantage in comparison with the developers was demonstrated several times in the case histories. The team of lawyers assembled by Chevron to fight the lone advocate from Citizens for a Better Environment over the oil refinery modification before the Bay Area Air Quality Management District Hearing Board is a case in point.

The necessity of a *consultation* process for siting new sources of air pullution has been stressed repeatedly. Royston (1979) concluded that: 'at least two factors are required to start a conflict, namely, the existence of a threat of pollution, and a population or interest group concerned about a particular environmental dimension' (p 60). The high level of local objection to new sources of pollution:

shows clearly the immediate, local nature of environmental concerns, and hence how essential it is for the enterprise to work with the local community, addressing local fears and concerns and meeting local needs. All too often this is not done ... (p 58).

Caldwell et al (1976) confirmed that, all too often, difficulties are generated by 'the perceived conflict between economic growth and air quality, especially from the viewpoint of industry' (p 136). They felt that citizen participation in decisions involving air pollution in a democracy is healthy, especially as the private sector frequently ignores its real social responsibilities in the pursuit of profit. The formal consultation processes of the land use control and, especially, the air pollution control agencies are generally unequal to the task in both countries. The non-statutory consultation exercise initiated by BECO in the North Carolina oil refinery case is, however, a notable attempt to consult the public thoroughly.

It is possible to advance a further indicator of the outcome of the siting process, which appears to hold true in the cases examined:

> The more personnel, financial, legal and technical resources objectors have, the greater their power, the more information and time they have, the more access to the decision process they have and the greater their support in the media, the greater their chance of stopping or seriously delaying the construction of a proposed new stationary source of air pollution or of securing additional mitigation of its effects.

THE AUTHORISATION PROCESS

The prevalence of bargaining in the authorisation process for a new air pollution source is apparent. Where sources are constructed, Downing and Hanf (1983b), in discussing the role of the developer, the control agency and the objector in reaching a decision, concluded that:

> The likelihood that all three groups would agree on the same environmental quality is small indeed. This generates conflict. Bargaining is a method of resolving conflict ... The result ... can eventually be an equilibrium where all three groups seem to agree that the achieved environmental quality is about right, even though, in the absence of costs and constraints, each would prefer a different outcome (p 333).

They believed that the nature of the institutions concerned had but little effect on the outcome of the siting process.

The outline in Figure 1 and the above discussion makes one thing abundantly clear: the more permits, whether discretionary or non-discretionary, that have to be obtained and the more permitting agencies that are involved, the greater is the chance of the developer's proposal being seriously delayed or stopped. The developer, after all, requires all the relevant permits to proceed, whereas the objectors only require one to be withheld or refused.

In one of the instances examined in this study, the developer in the Louisiana creosote storage plant case found himself faced by refusals of local and state land use permits, a postponement of the decision on the air pollution permit and a requirement to obtain a water pollution permit. He withdrew. Similarly, Dow withdrew its California chemical production facility proposal in the face of a large number of environmental requirements and the unusual refusal of an air pollution permit. When it appeared that Metro had obtained, or would obtain, all its permits to construct the Oregon energy recovery facility, it was defeated by another attack, through the initiative ballot process.

An obvious procedural corollary is that the more laws, regulations and rules which apply to the permitting activities of a given agency, the greater the chance that the proposal will run foul of one of them because the agency's scope for manoeuvre will be diminished. It thus appears to be true that:

> The more non-discretionary and discretionary permits that have to be obtained, the more rules and laws that have to be observed, the more permitting agencies that are involved, the more negotiations that are required and the more points of public access that exist in the permitting process, the smaller are the chances of a new stationary source of air pollution being constructed or of being built without very onerous control conditions.

IMPLEMENTATION OF CONTROLS

The stress on implementation in the air pollution control literature probably dates from Crenson's (1971) study of air pollution control in the neighbouring communities of East Chicago, Illinois, and Gary, Indiana[11]. Crenson's main contention was that powerful interests were capable, by the use of their power or even (as in his case of US Steel) simply by reputation, of reducing air pollution to a 'non-issue' by keeping it off the public agenda and hence preventing enforcement action. Since the time of Crenson's study, however, environmental consciousness has grown to such a degree that pollution is seldom a non-issue, especially when objectors have a formal opportunity to protest, as is usually the case when a new or expanded air pollution source is contemplated.

Once authorisation for a new or modified stationary source of air pollution has been granted, the degree of pollution arising from its operation may be less or, more usually, greater than the anticipated level. Thus, in the cases examined, the St Helens sulphuric acid plant emitted far more sulphur dioxide than expected and caused considerable damage. Similarly, the Florida resources recovery facility gave rise to much more odour and smoke than forecast. On the other hand, the Bolton lead battery plant has frequently emitted considerably less lead than the amount permitted.

Perhaps the most obvious reason for variations in expected pollution levels is the introduction of design changes between approval by the land use planning agency and construction. These can lead to emissions quite different from those considered appropriate by the planning agency[12]. Thus, once land use planning approval had been given, substantial changes to the design of the proposed Oregon energy recovery facility took place before it was aborted: a considerable reduction in capacity was agreed with the state Department of Environmental Quality and the type of air pollution control equipment was substantially modified.

Even where the conditions set by the land use planning and air pollution control agencies are met in the first instance, it is important to remember, with Brady and Bower (1982), that:

> Achieving original compliance is no necessary indication that continuing compliance will be achieved. The incentives inducing the former often have little impact in terms of achieving the latter.

Brady and Bower listed many of the technical reasons why a plant may not continue in compliance, including unanticipated changes in production process variables and the inadequate operation and maintenance of equipment. Other factors include the nature of the controls demanded, the acceptability of the controls to the developer, the enforcement of control by the air pollution control agency and the land use planning agency and the vigilance of the public.

Nature of conditions

Where the conditions set by the land use planning agency or the air pollution control agency require the installation of certain physical artefacts (eg, a chimney of specified height, an electrostatic precipitator, or an air filtration system) compliance can be easily checked by either agency and initial pollution problems should not normally arise. Thus, the Glossop molybdenum smelter and the new town glass fibre works were constructed with specified chimney heights and the Bolton lead battery plant was constructed with an air collection and filtration system to the satisfaction of the planning authorities. However, checks that the continuing operation of these types of equipment is within specification can realistically only be undertaken by the air pollution control agency, which should have the competent staff.

Emission control conditions that require direct monitoring or the checking of a company's own monitoring records will generally be the responsibility of the air pollution control agency because of its superior technical resources. Provided agency staffing is sufficient, conditions relating to the operation of equipment and emissions limitations can be checked relatively easily. Thus, it was easy to detect that the Florida resources recovery plant was frequently not complying with emission limitation conditions while it was operated by RRDC. Similarly, the pollution control authorities found it easy to check that the Bolton lead battery works was initially being operated within the set conditions.

Compliance with other types of condition (eg, requiring housekeeping and other general methods of pollution abatement or the avoidance of nuisance) may be more difficult to assess. The Maryland solvent recycling plant generally satisfied the few fixed emission limitation conditions relating to it but still gave rise to serious pollution.

Acceptability to developer

There can be little doubt that the degree to which there is compliance, with air pollution control conditions upon a new or modified source, will depend to a considerable extent on their acceptability to the developer. If he accedes to conditions only grudgingly as the price for his permission, his commitment to them may be low. Where, on the other hand, the developer freely negotiates the conditions with the air pollution control authority and the land use planning agency, he is likely to be much more committed to them. Thus, Chloride suggested its own lead emission limitation levels in the Bolton lead battery plant and has generally been proud to demonstrate compliance with them. On the other hand, conditions imposed on the Florida resources recovery plant by numerous agencies may have contributed to the negative attitude and non-compliance of the original developer of the facility. Certainly, conditions imposed by the St Helens council on the sulphuric acid plant, without significant consultation or negotiation, seem to have been disregarded at the outset of operations.

The *priority to pollution control* generally given by the developer will be a crucial determinant of the performance of any new plant. Most developers will be anxious to accord a level of attention avoiding the difficulties that beset such operations as the Maryland solvent recycling plant (which was closed down for a period) and the St Helens sulphuric acid works (which led to the developer being involved in considerable expenditure on legal expertise). Nevertheless, the range of developers' priorities exemplified by the case studies examined is considerable.

In the Forida resources recovery plant case the departure of RRDC and its replacement by a new operating company led to a marked improvement in the performance of the plant and a diminution in the level of complaint from local residents. Similarly, the takeover of the Cheshire fertiliser plant by UKF from the previous company led to significant reductions in pollution. Comparison between the attitude of Chloride in the Bolton lead battery plant case (voluntarily accepting low emissions) and Crewe Chemicals in the Yorkshire chemical formulation plant case offers convincing testimony of the importance of the priority accorded to pollution control by the developer. The Chloride case itself offers an example of the importance of management concern. The falling away of maintenance of pollution control equipment following a change of senior personnel led to a deterioration in emission levels which was rectified when the new managers reintegrated a maintenance schedule into the operation of the works.

Agency enforcement

Active implementation of controls by the *air pollution control agency* is known to have occurred in all but one of the cases where the source was constructed. Generally, air pollution control agencies prefer to discuss problems and agree solutions rather than take active sanctions. Nevertheless, the state of Maryland forced the closure of the solvent recycling plant, to enable it to re-equip, in the 1970s and it has subsequently continued to attempt to obtain full implementation of controls and improvement in performance. Such stringent application of the air pollution control agency's sanctions is very unusual. Evidence of more normal types of implementation of controls is provided by the state of Florida and Dade County in trying to enforce conditions to reduce pollution from the resources recovery plant and HMIP's efforts to improve the performance of Leathers Chemicals' sulphuric acid works. Fines were utilised in both of these cases and, over the years, the pollution control agencies do appear to have had some success in reducing air pollution levels in almost all the cases examined, though the new town glass fibre works continued to cause problems at the time of writing.

The *land use planning agency*'s powers of restrospective control are insubstantial in comparison with its anticipatory powers and with those of the air pollution control agency (Chapter 2). Further, the enforcement powers available to it are rather weak and there is a consequent reluctance to employ them. In addition, land use planning agencies are often unable to resist applications to vary planning conditions once a new development is operating (Chapters 3 and 5). These factors, coupled with a relative lack of technical competence in air pollution control, might lead to the expectation that they would take little interest in enforcing pollution control conditions or in trying to ameliorate serious pollution problems once sources have been constructed. In fact, in the case studies examined, the land use planning agencies were more active in implementation than might have been predicted. The reason for this is presumably the local nature of their jurisdiction and their readiness to try to respond to vocal manifestations of public opinion.

Of the ten instances examined where construction took place, active implementation was undertaken by the land use planning agency in six cases. For example, in the Maryland solvent recycling case it was Cecil County's action in harrying Spectron at the Providence plant and refusing discretionary permits at the Childs site that led to the agreement to construct a new facility. St Helens utilised first enforcement action and then a discontinuance order to try to close Leathers Chemicals' sulphuric acid plant. The known potential compensation cost of closure, several million pounds, showed the level of concern by the local authority. While the land use planning agency may not always be successful in attempting to implement its conditions or its general powers to control pollution from an existing source, it may thus sometimes achieve drastic remedies such as closure, as in the Texas asphalt plant case.

Vigilance of public

Vigorous complaint by the public about pollution from a newly constructed source is a crucial lever in ensuring active implementation of conditions. Thus, it was the complaints of the public about the Texas asphalt plant that led to its removal following land use planning agency action. Equally, the enforcement activities by Cecil County and the state of Maryland in the solvent recycling plant case flowed from public outcry. Again, the voluntary decision by Sterling Mouldings not to proceed with construction of the resin manufacturing plant in Tameside, once permission had been granted, was a direct consequence of the public protest about the pollution from its previous activities.

Overall, it is possible to advance an indicator of the nature of the implementation of controls over new or modified sources of air pollution which is true in the instances investigated:

> Once a new source of air pollution has been approved, the more closely the type of pollution control equipment is specified, the more the developer has been involved in negotiating the conditions, the higher the priority given to pollution control by the developer, the more the air pollution control and (particularly) the land use planning agencies insist on implementation of conditions and the more the public protests about pollution incidents, the greater will be the degree of control over pollution.

NOTES

1 See, for example, Van Meter and Van Horn (1975), Sabatier and Mazmanian (1981), and Fudge and Barrett (1981).

2 This requirement, of course, is fundamentally opposed to the British pragmatic approach in which policy is seldom defined by statute, but evolves frequently by 'laissez-faire'. See, for example, Griffith (1966:515–528).
3 See, for example, Janelle (1977).
4 The nature of the two processes is, of course, different.
5 Feldstein, M, quoted in Duerkson (1982:60).
6 Morell and Singer (1980) show that of 21 rejected or abandoned oil refineries on the east coast of the USA, 9 were rejected at the local level (p 189).
7 See, for example, DOE (1983).
8 On the whole, general public interest environmental groups prefer to concentrate their resources at the national level, to influence the making of policy (O'Riordan, 1979b) whereas very localised environmental groups often tend to oppose specific developments rather than consider broader strategies for shaping the environment (Lowe and Goyder, 1982:86–105).
9 There are numerous other ways of analysing the role of objectors. Dear and Long (1978) for example, have suggested that the alternative community strategies of exit, voice, resignation, illegal action and formal participation apply to locational conflicts.
10 The conflict of values involved in the weight that the believers in 'catastrophe or cornucopia' place on material or non-material goals is not easily settled by facts and rational argument. These groups have entirely different ideologies. See Cotgrove (1982).
11 Crenson applied Bachrach and Baratz's (1970:44) idea of non-decision making to the air pollution control process.
12 West and Foot (1975) described the construction cf a different kind of aluminium smelter from that discussed at a public inquiry in Anglesey. This gave rise to much greater emissions than forecast.

8
Comparing US and UK Preventive Pollution Controls

The United States and United Kingdom systems of anticipatory control over new or modified sources of atmospheric pollution are compared in this chapter, which is structured like Chapter 7. Conclusions about the nature of control in the two countries are drawn, again utilising the case histories to exemplify the observations made. Table 8 summarises the indicators derived in Chapter 7. These provide some guidance not only to the likely outcome of an application to site a new source in both countries, but to potential differences between the countries.

THE DEVELOPER

The outcome indicator points to two likely important differences in the role of the developer between the USA and the UK: the greater distrust of business in the USA might be expected to lead to more rejection of developers than in the UK; and the higher population density in the UK (and hence probable closer proximity of the source to homes) might be expected to result in less success than in the US.

Developer characteristics

Perhaps the most important difference between the USA and the UK in comparing attitudes to the developer is the much lower level of public distrust of business interests in the UK. In the United States the violation of environmental law carries much less of a stigma in the business world than it does in Britain. Business performance in the USA tends to be judged almost exclusively by economic rather than social criteria. As Vogel (1983a) put it:

> In a society without a Queen's Honors List, the only socially recognised measure of achievement in business is the 'bottom line'....
> While businessmen in America who 'stand up' to regulatory agencies – even to the extent of violating the law – are often regarded as heroes by other members of their industry, the opposite is true in Britain.

TABLE 8
Indicators of the outcome of the siting process
for a stationary source of air pollution

1 The more open the attitude of the developer, the greater the developer's local influence, the greater the benefits offered by the project, the more the developer is prepared to make meaningful concessions to the control agencies and local residents, the more industrial the previous use of the site and the surrounding area, the further the site from existing homes and community facilities and the less 'political' the pollutants, the better is the chance of the new source being constructed and operated.

2 Negotiations between the air pollution control agency and the developer may be protracted but will almost always result in the granting of a permit with the precise degree of control exercised varying, and being more stringent where there is significant external scrutiny, but characteristically taking little or no account of the geographic site of the air pollution source and hence of the actual damage likely to be caused in that location.

3 The land use planning agency will ultimately decide whether or not to allow a new source of air pollution to be constructed and, because its evaluation will frequently prove inconclusive (reflecting conflict between its objectives) it will usually seek to approve the source after negotiating increases in its benefits and reductions in its air pollution impacts, but will not use many of the planning control techniques at its disposal.

4 The more personnel, financial, legal and technical resources objectors have, the greater their power, the more information and time they have, the more access to the decision process they have and the greater their support in the media, the greater their chance of stopping or seriously delaying the construction of a proposed new stationary source of air pollution, or of securing additional mitigation of its effects.

5 The more non-discretionary and discretionary permits that have to be obtained, the more rules and laws that have to be observed, the more permitting agencies that are involved, the more negotiations that are required and the more points of public access that exist in the permitting process, the smaller are the chances of the new stationary source of air pollution being constructed or of being built without very onerous control conditions.

6 Once a new source of air pollution has been approved, the more closely the type of pollution control equipment is specified, the more the developer has been involved in negotiating the conditions, the higher the priority given to pollution control by the developer, the more the air pollution control and (particularly) the land use planning agency insist on implementation of conditions and the more the public protest about pollution incidents, the greater will be the degree of control over pollution.

Thus, in the case studies examined, marked opposition to developers in Britain did not generally arise unless or until the adverse environmental performance of their plants was apparent and their *reputation* had deteriorated. This was true, for example, in the St Helens sulphuric acid plant case and in the Yorkshire chemical formulation plant case. Only in the Bolton incinerator case did significant opposition arise prior to approval. In the United States, on the other hand, there was marked opposition to development in several of the cases studied (eg, the Louisiana creosote storage plant and the North Carolina oil refinery) before approvals were granted. In closer parallel to the British experience, there was opposition to the Florida resources recovery facility and to the Maryland solvent recycling plant only when these developments were in operation.

Contrary to the expectations aroused by the greater social responsibility of business in Britain and Vogel's (1983a) assertion about environmental concern, the developer's *attitude* to air pollution does not appear to have been markedly different in the US and the UK in several of the cases studied. Clear attempts by developers to meet their environmental responsibilities are perhaps seen most clearly in the Bolton lead battery plant and the North Carolina oil refinery cases. In both instances, companies made conspicuous efforts to meet not only legal requirements but to go considerably beyond them in accepting or proposing mitigating, non-mandatory, modifications and in adopting a co-operative posture. Equally, the attitude of four other companies, Leathers in the St Helens sulphuric acid works case, Crewe Chemicals in the Yorkshire chemical formulation plant case, Petroplex in the Texas asphalt plant case and Spectron in the Maryland solvent recycling facility case, showed little concern for their environmental responsibilities.

The balance of evidence from the other case studies, while not overwhelming, does tend to support the general expectation of the *influence* of British business. Thus, construction was authorised in seven, and took place in six, of the UK cases (Table 7), while permission was withheld in four of the US cases (Table 5).

Project characteristics

The American public probably reacts more vigorously to a political or a controversial *type of pollution* than the British do. Douglas and Wildavsky (1982) have commented on this phenomenon:

> Try to read a newspaper or news magazine, listen to radio, or watch television; on any day some alarm bells will be ringing. What are Americans afraid of? Nothing much, really, except the food they eat, the water they drink, the air they breathe, the land they live on and the energy they use ... How can we explain the sudden, widespread, across the board concern about environmental pollution and personal contamination that has arisen in the Western world in general and with particular force in the United States? (p 10).

Vogel (1986: 253–254) felt that the explanation for this phenomenon was widespread suspicion of business. Another reason may have to do with the inherent newsworthiness of the worryingly unprovable in a country where insecurity is part of the culture. Yet another may be that Americans have a greater belief that they can affect the outcome of decisions.

It is instructive that the opposition to the pollutants from the Dow chemical plant in California, to the dioxins from the energy recovery facility in Oregon, to the creosote from the storage plant in Louisiana and to the suspected carcinogens from the solvent recycling plant in Childs, Maryland, was particularly vigorous. By contrast, the opposition to the lead battery plant in Bolton was positively muted, though the campaign against the molybdenum smelter in Glossop used some American tactics in making emotive complaints about the dangers of heavy metal pollution.

The effect of the *nature of the site* on air pollution levels should generally favour the developer in the US because population densities are much lower than in the UK. It should therefore be easier to site a new source of air pollution well away from local residents to reduce the *likely damage*. Thus, the North Carolina oil refinery, the Texas asphalt plant and the Florida resources recovery plant were all at least half a mile from the nearest dwelling and the number of people resident within a quarter of a mile of several of the other American sources was very limited. In the UK, on the other hand, all the proposed sources were less than half a mile from the nearest houses. The problems posed by the St Helens sulphuric acid plant and the Tameside resin manufacturing plant were particularly exacerbated by the very close proximity of existing residential areas. However, the lack of immediate proximity to the source does not appear to have reduced the level of objection to proposals in the US and this factor does not seem to account for any important differences in outcomes between the two countries. Indeed, any British proximity factor tends to be more than counteracted by the long history of experience of industrialisation and the relatively low level of distrust of industry in relation to the USA.

Project benefits, particularly the effect on local revenues of a large development, in the USA can be substantial and local governments are frequently influenced by this. Thus, Solano County in the California chemical plant case, Brunswick County in the North Carolina oil refinery case and Oregon City in the energy recovery facility case were anxious to welcome these substantial developments. The importance attached to this revenue is exemplified by Oregon City extracting a condition that if Metro – which had tax-exempt status – rather than a private company, were to build and operate the plant then it would pay the equivalent of the local taxes to the city authorities.

Because of the different system of financing local government in the United Kingdom and the larger size of local authorities (Chapter 5) the revenue attractions of a large new source of air pollution are much smaller than in the

United States. It is comparatively rare for a UK developer of an industrial site to offer incentives to the local authority, other than the incidental benefit of the project in increasing or maintaining local employment. This factor alone can be highly significant, however, as the efforts of Bolton to ensure that Chloride did not construct its lead battery plant elsewhere testify.

Mitigation

The process of bargaining, which includes modification of the proposed project and the offering of concessions to the control agencies and to the local community, is perhaps more overt in the United States than in the United Kingdom because it is more public.

Despite his general *willingness to bargain*, the developer's relationship with the air pollution control agency in the United States is likely to be more cautious and restrained than in the United Kingdom. Despite all the rules (Chapter 3), there may be considerable uncertainty about the precise nature of the controls to be demanded in America, partly because changes in either federal or state rules during the permitting process may affect the outcome. Other reasons include the complexity of the rules, the varying interpretations placed upon them by the developer and the air pollution control agency and the degree of external scrutiny of the permitting process. In the United Kingdom, on the other hand, the general nature of best practicable means or local authority controls is well known and the relationship between the developer and the control agency is thus likely to be much less uncertain, though the precise details may need to be ironed out. The absence of external scrutiny may also mean that the air pollution control agency will be more overtly sympathetic to the developer's position than in the US.

Examples of the uncertainty in US requirements and consequent necessity to bargain can be seen in the Oregon energy recovery facility case. Here, stipulations of permissible levels of sulphur dioxide emissions were progressively lowered as more air pollution modelling was undertaken but hydrocarbon offset requirements were not, in the end, demanded when initially it had seemed that they would be. In the UK, the Chloride lead battery plant case in Bolton was an unusual instance of HMIP's air pollution control requirements changing as negotiations progressed.

The amount of information demanded of the developer by the air pollution control agency before striking a bargain differs markedly between the two countries. In the United States it can take a developer several months and the expenditure of large sums of money (eg, on ambient air quality monitoring) to assemble the information required for a permit application. Applications, which are subject to public scrutiny, are seldom deemed complete by agencies until yet more information is furnished. In the United Kingdom, on the other hand, the case studies confirm that the amount of information demanded is seldom great, is very unlikely to involve such intricacies as ambient air quality monitoring, and is not usually subject to external inspection.

There is some similarity in the nature of the developer's relationship with the land use planning agency in both countries, arising from the need for negotiation because of the uncertainty of the outcome. The seeming predictability of the US zoning system is illusory because of the requirements for various types of ancilliary permits and the opportunities they provide for delay or outright refusal. Similarly, though the development plans in the United Kingdom provide valuable guidance as to the likely acceptability of an industrial development, the decision is still taken essentially, as in the United States, on a case by case basis. Indeed, Johnson (1984) has remarked that, in general, the American land use planning system is a great deal more flexible than the British system and that very often, the persistent developer will eventually obtain his permit. However, this does not necessarily apply to new sources of air pollution, where deep public antagonism may change the outcome.

Thus, of the eight American cases examined, three land use planning agencies, St Tammany Parish in the Louisiana creosote storage plant case, Midland City in the Texas asphalt plant case and Cecil County at the Childs site in the Maryland solvent recycling plant case, refused permission to develop (Table 5). The North Carolina state coastal planning agency also initially refused to allow the Brunswick County plan to be altered to permit development but later accepted a revised plan. In Britain, on the other hand, only Bolton in the sewage sludge incinerator case refused permission for a new source initially, though Kirklees Borough Council began to issue refusals once pollution problems had started to occur in the Yorkshire chemical formulation plant case and Tameside temporarily issued a refusal in the resin moulding case.

The biggest difference in the bargaining relationship between the developer and the land use planning agency in the two countries probably concerns the appeal procedure. An American industrial developer appealing to hearing boards or to the courts in the states with the stronger land use planning control systems will have little chance of success if the correct procedures have been carried through – though he may have more in states with weaker controls. The local jurisdiction is free to make up its own mind irrespective of federal and state policy pressures. Where a state land use planning agency exists it is usually powerless to overrule a decision by the local jurisdiction. On the other hand, the UK developer – and the local planning authority – is aware that an appeal to the Department of the Environment is reasonably likely to prove successful, especially if the party in government is determined to encourage business. In Britain, therefore, a developer approaching a local authority knows that it must take account of central government policy, which will often tend to be overwhelmingly in his favour.

There was only one appeal to a non-local hearing board by the developer among the US cases studied, though there were several to the locally elected body which were all refused. In fact, the appeal to the Louisiana state Coastal Commission was refused in the creosote storage plant case, just as the appeal to St Tammany Parish Council had been denied. In the UK, four cases were the subject of appeal or similar inquiries at some stage: the Bolton incinerator, the Yorkshire chemical formulation plant extension, the Cheshire fertiliser plant extension and the St Helens sulphuric acid plant. Only in the incinerator case did the developer not gain his desired decision. Nevertheless, concessions to mitigate impacts are often offered by developers at public inquiries in the UK, as evidenced by the amelioration suggested in the Bolton incinerator case.

Notwithstanding the general impression of American developer/land use planning agency relationships, the developer of a new air pollution source in the United States will have less power than the planning agency. The developer's only recourse will often be to locate his development elsewhere if he cannot agree mitigation conditions or finds the *cost of control* too high. In the United Kingdom, on the other hand, the developer will have the greater power because of the likelihood of central government support on appeal but the local authority can use the delay endemic in the appeal procedure as a method of gaining concessions by threatening refusal. In the Bolton lead battery plant case, for example, Chloride was anxious to avoid the delays associated with an appeal against refusal of planning permission and was thus particularly amenable to bargaining.

The developer in the USA normally requires more land use permits than in the UK, where a single permit usually suffices. Different forms of mitigation may be demanded by each of the different local, state or federal agencies concerned. The small size of many local authorities in the United States means that the developer must pay more attention to the various project impacts on the immediate local community than in Britain. This is likely to lead to offers of more significant mitigation of the project's effects and of more substantial financial incentives to the local community in the form of planning gains. The payment for traffic management and road design work by Chevron in Richmond, California as a condition of modifying its oil refinery is a good example of the latter type of concession.

THE AIR POLLUTION CONTROL AGENCY

The outcome indicator would suggest that air pollution control conditions will be more stringent in the United States because there is greater external scrutiny of the process of granting the permit.

180 *Planning Pollution Prevention*

Agency characteristics

The system of air pollution control in the USA may be characterised as comprehensive, with each application being decided according to a clearly specified set of goals (ie, air pollution concentration standards) and to a state implementation plan designed to implement these standards. Though the detailed regulations are arcane (Chapter 3), air pollution controllers have a full range of anticipatory *powers* at their disposal to limit pollution from most new or modified sources. Their freedom of action is, however, more circumscribed by rules and scrutiny than in Britain.

In Britain, on the other hand, the *policy context* is quite different. There is no plan to reduce pollutant levels, each application is decided on an ad hoc basis and certain types of source are subject to relatively uncomplicated anticipatory controls typified by the best practicable means. Other sources are not subject to any air pollution control agency anticipatory controls at all.

Knoepfel and Weidner (1982) felt that Britain had only a 'national rump programme' of air pollution control. While it is quite obvious that the British system of pollution control is indeed incomplete in comparison with the American, their analysis seems to miss both the full extent and the sophistication of British controls over air pollution, particularly underrating HMIP's anticipatory controls. However, the partial nature of British air pollution control is well exemplified by the lack of anticipatory controls imposed by the air pollution control agencies in, for example, the new town fibre glass plant and the Yorkshire chemical formulation plant cases. The efforts of the local authorities to employ land use planning powers in these cases testifies to this lacuna. Lack of comprehensive anticipatory control is not a solely British characteristic, however. The way in which US companies can avoid meeting regulations is exemplified by the decision to locate the Petroplex asphalt plant exactly half a mile from the nearest house in order to gain a permit exemption.

The fragmentation of control over air pollution in the United Kingdom, with different types of industrial sources and mobile sources under the aegis of different agencies, is less likely to inconvenience an industrialist in pursuing his application than the division of functions between federal, state or regional (and sometimes local) agencies in the United States. Thus, in the United States the developers had to deal with both federal and state or regional air pollution control agencies in, for example, the North Carolina oil refinery and the California chemical plant cases. In addition, RRDC also had to deal with Dade County air pollution controllers in the Florida resources recovery plant case. In the UK, on the other hand, initially only Chloride in the Bolton lead battery works case had to deal with both HMIP and the local environmental health department, though HMIP also became involved in the later stages of the Yorkshire chemical formulation plant proceedings.

The degree of *external scrutiny* of the air pollution control process is much greater in the United States than in the United Kingdom. US air pollution control officials may be more bureaucratic than their British counterparts since they must be careful not only to follow the appropriate regulations but to be seen to do so. Thus, the air pollution controllers in the Maryland solvent recycling case were very anxious to be seen to be considering the permit application for the Childs site strictly according to the regulations. Those responsible for the North Carolina oil refinery permit quite openly admitted prolonging their examination because of the likely challenges from objectors.

Air pollution control conditions do indeed appear to be more stringent and onerous in the USA than in Britain, partly as a consequence of this scrutiny. It seems logical to equate the US 'best available control technology' with the British 'best practicable means' since both take the economic costs of control into account. However, there is no British equivalent of the use of 'lowest achievable emission rate' in non-attainment areas or of the complex monitoring conditions that are often utilised. For example, no British case involved conditions as onerous as those imposed by North Carolina on the BECO oil refinery which required several expensive ambient pollution monitoring programmes.

Project characteristics

The freedom of manoeuvre in interpreting the US rules according to the *nature of the site* is limited by the tightly prescribed policy context. It would not be possible, for example, for a major new source to be located in a non-attainment area without the provision of offsets. The Dow chemical plant case in California demonstrates the lack of flexibility to make decisions about major sources which might be locally desirable because of their revenue, employment and other benefits. This is not to say that the US air pollution agencies are completely without freedom of manoeuvre. Palmer Barge Lines were faced by a belated requirement to apply for an air pollution permit in the Louisiana creosote storage plant case, after initially being told no such permit would be necessary. The differences in particulate emission levels between the Florida resources recovery facility and the two Oregon recovery facility permissions (0.015–0.08 gr/ft^3) also demonstrate the variations in the degree of control that are possible in the US system.

Nevertheless, by comparison, the British HMIP may have considerably greater discretion in interpreting the meaning of best practicable means in relation to a particular new pollution source, if only because the absence of meaningful ambient standards means that prohibition is never necessary. The best practicable means for the general type of works may be publicly known but the history of the Inspectorate indicates the progressive and co-operative nature of the relationship between it and the developer once the source has been constructed.

The manoeuvrability inherent in British anticipatory controls is demonstrated by the Bolton lead battery plant case, where emission standards far less permissive than those normally required under best practicable means were agreed by HMIP. It may also be seen in the Glossop molybdenum smelter case and in the St Helens sulphuric acid works case where widely varying chimney height requirements were imposed once problems had arisen.

There appear to be few differences in the way air pollution control agencies deal with the *likely damage* arising from the precise location of a source in the US and the UK. In both countries it is left to the land use planning agency to state whether a given site is appropriate or not, though advice may sometimes be tendered by the air pollution control agency – especially by the environmental health department in Britain. Perhaps the most obvious difference is that some account of location may be taken by the air pollution control agency in the UK by prescribing higher chimneys where a source is in close proximity to housing or other sensitive land uses. The new town glass fibre works and the Glossop molybdenum smelter furnish examples of this method of control. Chimney height is not widely used in the US as a means of abatement, though some subtle tightening of conditions may occur where it is evident that a source is likely to cause damage to a nearby residential area. The additional activated carbon filter proposed for the Spectron plant at Childs, Maryland, following discussions with residents would be an example. In general, however, even less consideration will typically be paid to the precise location of the source in the US than in the UK.

Mitigation

The *cost of control* is obviously greater in the US, because of the greater stringency of abatement requirements. The recently encouraged flexibility in meeting pollution control objectives has been welcomed by American developers. An example of this is the use of bubble conditions in the California oil refinery modification case, to allow Chevron to choose its own methods of achieving the net emission decreases required.

Similarly, obtaining an air pollution permit is frequently a more time-consuming business in the United States than in the United Kingdom, where it is generally a brief process. There may be appeals and court cases to prolong the process in the US. It is noticeable, for example, that EPA took several months to make its PSD determination in all three cases where it was involved. The North Carolina air pollution control agency took ten months to issue a permit for the oil refinery. However, a consent will eventually be issued in the USA if the rules permit, just as it will in the United Kingdom. Despite the differences in legal provisions, negotiations to mitigate impacts are universal in both countries and it is almost unheard of for a serious applicant to be refused a permit.

Comparisons between the *personnel resources* of US and UK air pollution control agencies are difficult to draw, though the highly specialised British central government inspectors might be expected to be the most knowledgeable air pollution controllers. US air pollution controllers are, pro rata, much more numerous than their British equivalents. They are also subject to much more rapid rule changes and more frequent staff turnover than in the UK. Bargaining should therefore be easier in Britain.

The relationship between the local air pollution control agency and the land use planning agency is likely to be closer, and *consultation* between these agencies is likely to be greater, in the UK than in the US because they operate at the same level of government and because their decisions are both usually taken on a case-by-case basis. It is quite common for environmental health department officers in English districts to ask their colleagues in planning departments to append conditions to mitigate pollution. This is seen, for example, in the cases of the new town glass fibre works and the Yorkshire chemical formulation plant. That such consultation does not always take place is evident from the St Helens sulphuric acid plant case: the Glossop molybdenum smelter case shows that it sometimes results in disagreements between the two local authority agencies. The somewhat distant relationship between HMIP and the local planning authorities in the UK is probably more akin to the US situation.

THE LAND USE PLANNING AGENCY

The biggest difference between the performance of the land use planning agency in the US and the UK to be surmised from the outcome indicator is probably the likelihood of greater variability in decisions in the US because of the small-scale, responsive nature of local government and lack of centralised control there. Thus, on the one hand, more refusals and more stringent conditions might be expected in the US while, on the other hand, greater capitulation to the interests of industry might also be expected, according to local circumstances.

Hall (1982) has probably offered the most succinct comparison between the local government (ie, land use planning agency) responses to environmental issues in the United States and Britain:

> The American planning system ... is much more diverse, much more localised, much more multi-centered than its European counterpart. The hand of central (Federal or State) government is less evident ... politics are more transactional, more committed to wheeler-dealering and trading of issues and votes, than would normally be the case in Europe. The community interests tend to be better defined and better organized, ... because of the traditional diversity of the country ... The media ... are more concerned with local issues ... All this adds up to a much more free-wheeling, rapidly-shifting, diversified pattern of politics than is normally seen in Britain (p xxi).

Agency characteristics

At first glance, the British and American approaches to land use planning are the reverse of their approaches to air pollution control. Thus, while the United States has a comprehensive national air pollution control system in which decisions on particular applications are used to implement statutory plans, its approach to land use control is partial, fragmented and very variable in different parts of the country. The United Kingdom approach to air pollution control is similar – partial and fragmented – but its land use planning system is a comprehensive national one in which decisions on applications are used to implement statutory plans.

Johnson (1984) has explained some of the reasons for the differences of approach between the two countries:

> As a nation with the historic task of developing a frontier, the United States evolved an individualist ethic... Thus its planning ethos favours growth... Land is expected to be developed to its 'highest and best use'...
> British planning practice has been cordial to private development also, but with the sense of the 'frontier' more remote in its history... Britain's view of the public interest is more communitarian than the American, valuing public goods that transcend the aims of particular individuals and businesses.

The growth of suburbanisation within a clear urban-rural boundary in Britain and the ubiquitous suburban sprawl in the United States are apparent to any observer. This difference offers convincing testimony as to the relative *powers* of the planning systems to control the location of new development. This, in turn, reflects the basic objectives of the two systems:

> Those of the British land use planning system, ... are to preserve open land, preserve agriculture and contain urban growth .... Land use planning and control in the United States are intended to guide development, which is not merely accepted but sought (Clawson and Hall, 1973:168).

While the American system may not be able to prevent development altogether, it can control the kind of development that takes place (Hall, 1977).

Haar (1964) characterised the differences between the planning systems as follows:

> [In Britain] the essence of planning control is that all development – broadly defined as any material change of use – requires official permission. The American system of land-use controls differs in that it is largely voluntary and, with its maze of statutes, structurally far more complex... and far less centralised as to its sources (p 254).

Britain possesses, as Garner and Callies (1972) put it: 'a single coherent system, whereas the American is perhaps scarcely a system at all'. Indeed, Heidenheimer et al (1975) said that the United States is a nation where 'town planning starts not with a plan, but with a mortgage' (p 121).

Delafons (1969) believed that:

> The American and British approaches to the problem of controlling private development represent almost the opposite extremes in planning methods. But the distinction between a formal system of regulatory controls, which eschews discretion as far as possible, and the alternative of exercising control as a discretionary power in government is perhaps more apparent than real (p 112).

Haar (1984) also stressed the similarities between the two countries: 'the country most similar to the United States in its urban planning system, both functionally and historically, is Britain' (p 254).

In practice, therefore, the land use planning systems are not quite so different as they seem. The US planning system has more powers than at first appear obvious and, as explained in Chapter 3, the range of these has grown substantially in recent years. Indeed, the advent of cluster zoning and planned urban developments in the United States that require individual permits and of simplified planning zones in the United Kingdom indicate that the two systems are rapidly converging.

In the United States there is far greater variation, in both plan-making and zoning activity, between different authorities than in the United Kingdom. There is also far more complexity. Though the extent of what the British term permitted development is much wider in the United States, in parts of some states at least 30 sets of development regulations may apply to some developments, even if the sites concerned are properly classified under the state land use law and appropriately zoned. Further, there are the various federal laws:

> Well-intentioned as they are, [the federal laws] add yet another series of land use regulations that restrict the use of land, a series of regulations that is difficult to coordinate, much less prune or delete (Callies, 1984:171–172).

The variation in the *policy context* of land use control in the USA is exemplified by the weak systems in Louisiana and Texas (where county zoning is forbidden) and the stronger ones in Oregon and California. The lack of dependence of American local authorities on financial aid from the state or federal treasury and the fact that appeal is generally to the courts, rather than to central government, means that there is far less central control over the land use decisions made by US agencies.

It is inevitable that an application to construct or extend a new air pollution source will involve conflict between planning objectives. These may be most acute in the British planning system, where there is less apparent certainty and stronger adherence to planning policies. However, they also arise in the United States, not least between the objectives of various agencies which may have to issue land use permits under conflicting legislation. It is significant that the only British refusal of a permit stemmed from conflict with policies embodied in an informal local plan (the Bolton incinerator case) and that the three refusals by American local land use planning agencies were all made in the absence of a strong relevant policy context. These applications, for the Louisiana creosote storage plant, the Texas asphalt plant and the Maryland solvent recycling plant, could all have been granted without obvious conflict with the local policies in force when the applications were submitted.

The comprehensive UK land use planning system presupposes that all decisions about new sources of air pollution will be taken within the context of the statutory development plan, whereas the partial US system is based upon a variable array of land use controls and plan making is usually completely divorced from the control of development. While many British plans do contain policies relating to air pollution control, this is comparatively rare in the United States. In general, there was little indication that the pollution control policy context was specified sufficiently clearly in land use plans to allow land use planning agencies to act in accordance with it. Decisions on new air pollution sources thus normally have to be taken on an ad hoc basis.

The parochialism, the independence, the openness to *external scrutiny*, and the susceptibility to political pressure of US land use planning agencies make elected representatives and appointees more likely to resist a new stationary source of air pollution than their British counterparts, though the incentives offered by the developer can have a disproportionate effect on the outcome. There is more information available on most applications in the US than in the UK, and consequently often more informed and purposeful public pressure to refuse an application.

American land use planning agencies often have the discretion to refuse polluting developments, even though the zoning may be appropriate. Thus, the agencies in, for example, the Maryland and Oregon cases could equally have accepted or refused the developments concerned. In fact, the local Maryland agency refused the solvent recycling plant largely on procedural grounds, whereas Oregon City accepted the energy recovery facility.

An additional factor militating against construction of such sources in the United States is the number of agencies involved in granting land use permits apart from the local government responsible for zoning, since this multiplicity increases the opportunities for delay or refusal. Morell and Magorian (1982) have emphasised the problems of 'double-veto' by state and local agencies:

In most states local land-use regulatory authorities, cities and counties, can veto the location of a new facility by withholding the necessary discretionary zoning or other land-use approval .... This double veto usually precludes siting new facilities because local authorities would not balance local desires versus statewide interest in both environmental quality and industrial management (p 91).

In the eight cases examined in the United States, there were three refusals at the local level (the Maryland solvent recycling plant, the Texas asphalt plant and the Louisiana creosote storage plant) and two refusals at the state level (the Louisiana coastal use permit and the North Carolina coastal plan – later reversed). The eight UK cases included only one refusal – the Bolton sewage sludge incinerator.

Project characteristics

It is apparent that the relatively unpoliticised British land use planning agencies are likely to be more active and influential in shaping the outcome of decisions than their US counterparts. As Heidenheimer et al (1975) stated:

since a good part of the power, both political and financial, stems from their relations with national agencies, [British] local officials are not so totally dependent on the local political situation, and they have more manoeuvrability than does the American city executive in negotiating with local interest groups (p 115).

It is nevertheless the case that, despite the greater range of anticipatory powers available to American air pollution controllers, land use planning agencies have to make the final decision to grant or refuse approval for a new source according to its costs and benefits and the *nature of the site*.

The greater dependence of local authorities on local taxes – and hence on new industry – in the United States, because of the much lower proportion of funding from central government direct grants, would be expected to encourage the local land use planning agencies to give permissions more readily than in the United Kingdom. There is indeed evidence from this study that it is the smaller, rather than the more remunerative larger developments, which are refused in the United States. US local authorities with a substantial existing tax base will not be greatly influenced by the prospect of further but proportionately insignificant property tax revenue. Conversely, the poorer authorities may be particularly anxious to secure substantial new developments.

In the US case studies examined, the variety of responses ranged from outright rejection through the imposition of detailed air pollution control conditions to rapid unconditional acceptance (eg, Solano County's response to the Dow Chemicals proposal in California). In the UK cases the responses also varied but overall there was much more uniformity of decision than in the US. Thus, in seven of the eight British case studies planning permission was granted, or would have been granted, with conditions relating to pollution control. The costs of the decision are overtly considered by land use planning agencies in both countries: projects offering major benefits are seldom refused. Nevertheless, the differences in refusal rate, and in responses to major developers confirm the expectation, from the outcome indicator, of decision variability.

Land use planning agencies in both the US and the UK used refusal of permission to prohibit emissions. British planners also placed more faith in the use of high chimneys to disperse pollutants and reduce *likely damage* than their US counterparts. For example, chimney height conditions were used in the Bolton incinerator, the Glossop molybdenum smelter and the new town glass fibre works cases.

Mitigation

The mitigation of the impacts of the development during the land use control permitting process is often enshrined in planning conditions or agreements of various kinds. Two factors militate towards the greater use of planning conditions relating to the control of air pollution in Britain than in the United States. The first is the more comprehensive nature of the US air pollution control system and hence the lack of need for planning conditions. The second is the more comprehensive nature of the British planning system and hence the ability to consider and utilise conditions limiting air pollution. In the US cases, Oregon City was the only local agency employing planning conditions – imposing nine on the energy recovery facility. (The Florida state power plant siting agency also used such conditions in the resources recovery case.) The local authorities in the British cases imposed conditions relating to air pollution on every one of their planning permissions though the central government deleted them in the Cheshire fertiliser extension case.

The *cost of control* is clearly an important issue in mitigation, and these are not always associated only with the air pollution source. The negotiation of concessions of a non-environmental nature is perhaps more common in the United States than in Britain. Thus, the agreement of Chevron to contribute towards road design costs in California and of Metro to pay the equivalent of property taxes to Oregon City appear to have had no parallel in the British case studies. Generally, it may be more onerous to meet the conditions imposed by land use planning agencies in the US than in Britain, because of the costs involved.

Fix and Muller (1982) confirmed the highly political role of local governments in increasing the costs of environmental regulation in the United States:

> Regulatory costs are only infrequently and selectively attributable to state and federal rules.... Rather, regulatory costs reflect the attitudes of the local population toward growth which is reflected through the political process (p 3).

While British planning authorities do seek similar planning gains, their attempts are usually confined to commercial development and do not normally extend to industrial air pollution sources.

British planning authorities, with at least 20 years more experience of land use controls behind them than their US counterparts, tend to have better financial and *personnel resources*. They employ more professional planners and are held in higher regard than US planning agencies. According to Scarrow (1971) this difference in esteem stems from:

> Widely held cultural values, such as commitment to land-use planning; the influential role of public officialdom and the acceptance of that role; and a standard of public conduct which Americans may rightfully envy.

This difference of view about the reliability of local government officials led to the setting up of the complex US decision machinery and the search for clear decision criteria in zoning and ordinances, as Heidenheimer et al (1975) explained:

> The suspicion of municipal officials, born in the machine era of American politics, led the reformers to create a multitude of citizens' boards and commissions, whose separation from city hall was intended to guarantee that public planning could be carried on in isolation from the corrupting influences of party politics. The proliferation of such commissions ... produced fragmentation and deadlock (p 113).

Local government in Britain, by contrast, is generally trusted to take decisions:

> Policy making tends to be carried on intra-governmentally, by elected or appointed officials, rather than in extra-governmental settings (Heidenheimer et al, 1975: 115).

Further, in contrast to the pluralist pressures on US planning agencies, in Britain:

> Planners do not experience the same degree of pressure from either powerful private interest groups or citizen lobbies that American planners must continually confront.... the European city planner is typically less responsive to external pressures (Heidenheimer et al, 1975: 120).

It is perhaps not surprising that the quality of American planning staff is sometimes low and turnover tends to be high. This must be another factor contributing to the less consistent approach to decision-making on proposals for polluting development in the USA.

There was little discernible difference between the attitudes and knowledge of planners about pollution in the two countries. Planners generally tend to take a rather neutral role in decisions involving air pollution. Thus, despite the more professional and active nature of British planning authorities, any conditions they may impose are usually added at the request of environmental health officers: as in the Bolton lead battery plant case and the new town glass fibre plant case. There are exceptions, which depend to some extent on whether in-house advice is available. For example, the planning officers in Cheshire – who had no other county expertise to call on – were actively concerned with air pollution control in the fertiliser plant extension case. Cheshire County Council hired consultants to provide expertise; as did Oregon City, in framing its conditions. Nevertheless, UK land use planning authorities are probably more likely to take an interest in pollution matters without being prompted by objectors than their US equivalents.

In each instance where either an EIA or a British inquiry took place environmental problems appear to have been more clearly identified with benefit to their subsequent mitigation. There were four American examples of the use of environmental impact assessment: the North Carolina oil refinery; the California oil refinery modification; the California chemical plant; the Florida resource recovery plant, and no British examples. However, the information revealed in the two British cases where a public inquiry into the granting of a permission was involved (the Cheshire fertiliser plant extension) the Bolton incinerator and was significant, if less comprehensive in scope. The US EIA documentation may, by providing information about likely impacts from major developments, have supplied more ammunition for project opponents and led them to modify their tactics, but this did not cause projects to be refused by land use planning agencies in the cases concerned.

There were few differences discernible between the use of land use control techniques such as detailed site design, buffer zones, siting with respect to terrain, or siting with respect to sensitive land uses (Chapter 2) in the case studies in the two countries. Such use was severely limited (Chapter 7), though the negotiated relocation of the Maryland solvent recycling plant did provide one American example of removal of a pollution source from a valley and its reconstruction on a flat site in an industrial area.

There is more *consultation* and co-operation between planners and pollution controllers in the UK than in the USA, though there is also more conflict. In Louisiana and Texas for example, there is rarely any contact between the two types of agency. In the case study examples, there was co-operation in Dade County, Florida and in the EIA process in the California oil refinery modification case. None of the administrative and legal tensions between local land use planning agencies and pollution control agencies so evident in Britain in the cases involving planning conditions on works controlled by HMIP (for example, the St Helens sulphuric acid works and the Bolton lead battery works) seem to arise in the United States. Even in the Oregon energy recovery facility case, where planning conditions were imposed by the local land use planning agency, there was no dispute between the control agencies. This may well be because overlapping jurisdictions are endemic in the USA, where the primacy of the air pollution control agency is clearer.

THE OBJECTORS

Almost all the factors in the outcome indicator would suggest that objectors will have a greater chance of influencing decisions on new sources of air pollution in the United States than in Britain.

Objector characteristics

By having the ability to cause delay by the use of appeal procedures involving the courts, US objectors have greater effective *sanctions* against the developer of a new stationary pollution source than objectors in the UK. It was this sanction which BECO was anxious to avoid by making numerous environmental concessions in the North Carolina oil refinery case.

It is, however, the difference in *access* to the decision-making process in the two countries which provides the most powerful explanation of the variations in outcome. As Brickman et al (1985) stated:

> The effectiveness of private intervention in regulatory proceedings depends in large part on access not only to the decision maker but also to relevant sources of information. The US regulatory process is notable for exposing such information to public scrutiny.... In contrast, European governments continue to carry out their regulatory business under a presumption of confidentiality. The edifice of secrecy is more firmly grounded in Britain (p 44).

Johnson (1984) observed that the American political culture appears to sanction and motivate citizen participation more than does the British. Citizens have a constitutional right to participate and tend, accordingly, to pursue their own interests unashamedly. In Great Britain by contrast, public participation tends to be justified on broader political and administrative grounds, and is ostensibly geared to 'good planning' for the whole community.

There are thus many more opportunities for public participation in the United States than in the United Kingdom. Enloe (1975) felt that the growth of the environmental movement in the United States owed much to these opportunities:

The analysis of environmental politics demonstrates that citizen access to scientific expertise and governmental fragmentation providing a multitude of alternative access points both enhance the launching of an effective citizen movement (p 323).

The role of objectors in regard to the grant of a permit by the air pollution control agency is markedly different in the UK and the US. In the United Kingdom, there is no public or political involvement in the decisions of HMIP, which have never been challenged in the courts. Similarly, there is virtually no political, and no public, involvement in the decisions of environmental health departments in relation to new sources: decisions are taken by salaried officials. In the United States, on the other hand, there are provisions for open access to air pollution control agency files relating to new sources and there are usually provisions for public hearings, and for appeals against the agency decisions by the developer and – sometimes – by third parties. It is not uncommon for further recourse to be taken to the courts.

While it is unheard of for objectors to cause the air pollution control agency to refuse its permit – even in the California chemical works refusal, the objectors' role was encouraging rather than causative – it is quite common for them to appeal. Thus, the Texas asphalt plant case and the California oil refinery modification case provide examples of formal appeals by objectors against the grant of air pollution permits.

While both the National Commission on Air Quality (1981) in the US and the Royal Commission on Environmental Pollution in the UK (1976) have criticised the respective arrangements for public participation, there is no doubt that the formal rights of the public to be involved and to take legal action are much greater in the US.

There are generally ample opportunities for objectors to participate in the land use plan preparation process in both countries. However, the role of objectors in the grant of a permit by the land use planning agency differs between the US and the UK. In the United States, objectors have the right to see virtually all the information held by the planning and zoning department of a local jurisdiction, have the right to object in writing, have the right to be present at many meetings and have the right to speak at a public hearing for a prescribed period of time. This type of participation occurred in many of the US case studies. For example, there can be no doubt as to the influence of objectors in the decisions to refuse the Louisiana creosote storage plant, the Maryland Childs solvent recycling plant and the Texas asphalt plant, or in determining the conditions in the Oregon energy recovery facility case.

In the United Kingdom, though a register of applications is kept, applications are often advertised or notified, and objectors have the right to see the planning application and committee reports and to write to the authority to object, there is no objector right to see other relevant material (the comments of those consulted, for example) and no formal right to be heard at the committee or council meeting at which the decision is taken. A hearing of objectors is sometimes permitted, as in the Tameside resin manufacturing plant case, but this has not been common practice. In the UK, as in the US, controversial decisions will often be taken by elected politicians rather than by salaried officials, and they are answerable, at least in principle, to the objectors and other members of the electorate at the ballot box.

Third parties in the United States may object to a planning decision and have the right to be heard according to local circumstances, by the elected body if the decision was originally taken by a nominated body and to appeal to the courts. In the cases studied, the objectors appealed to the council of Oregon City and to the Land Use Board of Appeals in the energy recovery facility case and to the courts against Solano County in the California chemical manufacturing plant case. These opportunities for public participation in the land use decision making process are not limited to the local level, as the state hearings in the North Carolina oil refinery and the Louisiana creosote storage plant cases illustrate.

British objectors have no third party right of appeal unless they can prove that some legal requirement was not met; when their appeal would be to the courts. However, should the developer appeal against a refusal or the nature of conditions imposed, then the objectors may take as long as they wish to present their case at a public inquiry and they may cross-examine the witnesses called by the developer. The appeal process in the UK, therefore, gives the objectors a potentially excellent platform from which to make their case. They may wield considerable influence at these inquiries by winning environmental concessions from the developer; as in the Bolton incinerator case.

Because local authorities are generally bigger in the United Kingdom than in the USA, the proportion of their resident population suffering the effects of pollution from a new source is likely to be smaller than would be the case in America. Liaison with legislators will also tend to be less close in Britain. Similarly, the benefits of investment in new facilities, or the expansion of existing ones, accruing to each resident are likely to be smaller in the UK. Thus, public access to the British decision-making process does not guarantee that environmental objections will triumph over other considerations. As Vogel (1986) stated:

> In the case of pollution control, the lack of opportunities for public participation may mean that less weight is given to environmental considerations; in the case of the planning system, it is often precisely the relative accountability of local planning authorities to public pressures that undermines the political influence of amenity interests (p 127).

In the case studies, the four US examples of land use permit refusals contrast with only one in Britain. There was no British equivalent of the Oregon energy recovery facility case, in which an initiative ballot was won by the objectors to defeat the project after the land use permit had been granted. Overall, therefore, objectors thus have a considerably better chance of preventing the construction (and, to a lesser extent, the expansion) of a new stationary source of air pollution through the land use control process in the US than in the UK.

Another factor in the relative success of American objectors' campaigns is the proliferation of local television and radio stations and of local newspapers. Skilful use of these widespread media opportunities ensures that the views of opponents to a new source in the United States receive wider coverage than in the United Kingdom. The *media attention* provided by the *Oregonian*, in the energy recovery case, and the *Wilmington Star*, in the North Carolina oil refinery case, had a considerable effect in alerting and mobilising public opinion. The objectors in the Texas asphalt plant case became locally celebrated as a result of their numerous appearances on the television. There is no equivalent to these cases of high profile mobilisation of media support among the British examples studied.

Project characteristics

While the *nature of the sites* had a considerable influence on the outcome of the case studies, the size and composition of the population affected by the sources was not an important factor in explaining the difference of objectors' responses in the two countries. Similarly, the nature of the argument in objectors' cases varied considerably in both countries, though there was, perhaps, a greater tendency to exaggerate the *likely damage* from certain political pollutants, especially those associated with well publicised health hazards, in the USA. Such pollutants tend to arouse more public hostility in the USA where there is a greater tendency to assume that some pollutants are 'unsafe at any concentration'.

Mitigation

The extent to which objectors are able to secure additional mitigation of the impacts of a new air pollution source depends not only on their access to the decision making process. In the USA environmental pressure groups and other objector groupings tend to have more finance, more personnel, more legal and technical *resources*, more time and more information than their counterparts in the United Kingdom[1]. They are thus able to mount thorough, professional, well-argued opposition to a wider range of new pollution sources than in the UK (Symonds, 1982). The Citizens for a Better Environment group, which was active in eliciting additional mitigation in the California oil refinery modification case, is a good example of a specialised organisation involved in air pollution issues at the local level without a UK equivalent. O'Riordan (1979b) reported that, though British groups are ill-staffed and ill-equipped compared with their American counterparts, they can be surprisingly effective because of the case-by-case, discretionary and consultative nature of the system.

The heavy reliance on legal debate and process in the American system is illustrated in several of the case studies. For example, the objectors were represented by a lawyer before the Air Control Board in the Texas asphalt plant case. In the California oil refinery modification case Chevron protested through the courts that Citizens for a Better Environment had no right to appeal and then used four lawyers to help defeat that appeal. The superior organisation of the US objectors in using the courts, and mobilising opinion, and the general professionalism of tactics was evident, for example, in the commissioning of a film in the North Carolina oil refinery case and in securing a ballot in the Oregon energy recovery facility case.

US objectors are able to use the *consultation* process to mitigate the impacts of a new development to a greater extent than their British counterparts. In the case studies, objectors won concessions on several air pollution permit conditions in the United States. Examples include the activated carbon filter in the Maryland solvent recycling case and the increased percentage of sulphur dioxide removal in the Oregon energy recovery facility case. In view of the paucity of opportunity for public involvement outlined above it is not surprising that such concessions are much less common in Britain. None of the British case histories provides evidence of an air pollution control agency tightening its conditions at the behest of objectors before construction commenced.

As well as facilitating public access to nearly all non-commercially confidential files, US agency officials are prepared to discuss the merits of an individual case before a decision has been taken. Indeed, they frequently welcome pressure from public interest groups because they feel it can result in the justification of a permit in which the conditions are less biased towards the needs of industry. An example of such positive involvement, albeit manifested in an adversarial manner, would be the role of Citizens for a Better Environment in the California oil refinery modification case, where the nature of the conditions changed from 'no net increase' to requirements for a 'net decrease' in emissions. This willingness to discuss particular cases prior to approval in any other than the most general terms – and then only rarely – is conspicuously absent in Britain.

Objectors in the USA also have a better chance of winning meaningful concessions in the form of mitigation measures in the normal land use control process. The tightening of conditions at the behest of objectors in the Oregon energy recovery facility case provides a good example. However, the public inquiry system in Britain can provide similar opportunities for mitigation, as the Bolton incinerator case demonstrates.

THE AUTHORISATION PROCESS

The outcome indicator, and virtually all the factors discussed above, suggests that the developer of a new stationary source of air pollution will have a greater chance of constructing the plant in the United Kingdom than in the United States, irrespective of the environmental merits of the proposal. If constructed, a source would be expected to be subject to more stringent controls in the US than in the UK. Corporate or economic interests will generally prove more powerful than environmental interests in the UK; but not necessarily in the United States, except where modifications of existing major plants are concerned.

There appear to be surprising similiarities between the two apparently very different control systems. Notwithstanding the adversarial, public, standard-based approach to regulation in the US and the co-operative, private, ad hoc approach in the UK, the universality of bargaining and of the tendency to grant air pollution permits to serious applicants are clearly demonstrated by the case study histories presented. The importance of land use planning agencies in taking decisions affecting local air pollution levels is obvious in both countries. Further, the two systems appear to be converging. This convergence is evidenced by the British adoption, as a member of the European Economic Community, of certain air pollution standards; and by the increasing use of co-operative mechanisms, such as one-stop permitting, emissions banking and discretionary emission reductions in the USA. In land use planning control, the American use of flexible zoning instruments like planned unit developments and the British interest in simplified planning zones testify to this movement towards a 'mid-Atlantic' regulatory system with a number of common features.

Nevertheless, the authorisation process for a new source of air pollution involves several elements which differ significantly between the UK and the US. Because there is less trust of public servants, politicians and business in the US than in the UK, the rules of government tend to be written in far more detail and consequently there are more environmental laws and rules to be observed, and permits to be obtained. For major developments in the US, it is not uncommon for over a dozen different environmental permissions to be required from federal, state and local government and for public participation and appeal procedures to apply to many of these, frequently involving the use of the courts. However, with the exception of discretionary land use planning permits, the effect of these regulations tends to be to delay, and possibly eventually to force withdrawal from, a project rather than to stop it outright. In the United Kingdom planning permission is normally the only significant environmental hurdle to be cleared; though this can involve a multitude of ameliorative requirements.

The North Carolina oil refinery needed over ten permits and, notwithstanding the attempts of state officials to co-ordinate the granting of these and to keep track of the application, BECO was hard-pressed to maintain its schedule through the regulatory process. Similarly, numerous federal, state and local permits were required in both the Florida resources recovery plant case and in the California chemical works case. In the eight United States case studies, only the latter application foundered on the environmental regulations of other than a land use planning agency. Even here it is likely that Dow Chemicals would have received its air pollution permit if it had waited for the offset provision rules to be changed. In the UK case studies there was not even evidence of any delay in gaining permissions being engendered by air pollution control regulations.

There is much greater resort to the courts in the United States than in the United Kingdom in the siting of new air pollution sources and in environmental regulation generally. Asimow (1983) has contended that Britain is a nation of remarkably few laws, rules and regulations and that many of the latter relate to more trivial matters than in America. He argued that, while part of the reason is the greater tendency in the UK to write controversial requirements into the statutes rather than into rules:

> The most significant explanation seems to flow from the deep cultural differences between the countries. Americans traditionally distrust officials and favour advisory procedures and judicial interventionalism. In Britain, on the other hand, people are comfortable in relying on official discretion to strike compromises and make individualised judgements which are never reviewed by the courts.

He felt that political opposition to the substance of a government decision is directed against the procedure by which it is taken, citing the controversy over rule-making in the US and over major public inquiries in the UK.

Kelman (1981), in discussing the paradox of American self-assertion and the propensity to settle disputes not by compromise but by appeal to a third party, the court, stated:

> How strange it is that in contemporary America, the society whose democratic roots go deepest, the courts, that least democratic branch of government, should have more power than in any other Western democracy (p 236).

Wilson (1985) rejected the notion of culture as an explanation for the differences in Anglo-American regulatory approaches. Rather he adopted the explanation of 'political choice' in the creation of agencies, seeing the US as 'pluralist' and the UK as 'corporatist'. However, the conflict, interest-group self-seeking and resort to the courts which are virtually endemic in the pluralist USA probably arise from culturally determined political choices. There is evidence that, here too, British and American approaches are converging.

Only in one British case study, the St Helens sulphuric acid manufacturing plant, was there court action; and this was over an enforcement issue. On the other hand, the courts were involved in the Maryland solvent recycling plant case (to such an extent that the developer qualified as a lawyer in order to conduct the legal arguments), the Texas asphalt plant case, the California oil refinery modification case and the California chemical plant case. Quasi-judicial appeal procedures were also involved in the Louisiana creosote storage plant, North Carolina oil refinery and Oregon energy recovery plant cases.

IMPLEMENTATION OF CONTROLS

The outcome indicator tends to suggest that there might be little difference in the extent to which pollution controls will be implemented in the USA and Britain. From the previous discussion, it might be expected that, while the British developer would give greater priority to pollution control, the American public might protest more about pollution.

Nature of controls

Conditions applied to new sources of air pollution tend to be more complex and onerous in nature in the United States than in Britain. This might be expected to lead to greater difficulties in ensuring compliance but the case studies do not provide sufficient evidence to support this contention. Breaches of conditions were monitored in both countries but there were difficulties in enforcement in both the US and the UK, partly because it was not always possible to pinpoint the reason for exceeding the conditions.

Acceptability to developer

It is possible that the effectiveness of air pollution control conditions will prove greater in the UK than in the US because the majority of UK companies probably show a higher degree of environmental responsibility than their American counterparts and thus are motivated to achieve a greater degree of co-operation with the air pollution control agencies. Planning conditions relating to air pollution control may also prove more effective in the UK, for similar reasons.

The attitude of developers to non-compliance in the United States appears to be more relaxed than in Britain, notwithstanding the relative number and size of the fines imposed. However, in the case studies, the acceptability of conditions to the developer did not appear to be markedly different in the two countries. In practice, the attitude of the developers in the case studies did not vary much either, though there was much more coercion in the United States enforcement cases. This can be seen by comparing, say, the Bolton lead battery works case with the Florida resources recovery plant case.

It appears that the *priority to pollution control* accorded by developers is very dependent upon the pressure applied by the control agencies and/or upon the degree of protest from local residents. It thus varies more substantially within the two countries than it does between them.

Agency enforcement

The *air pollution control agency* in both the US and the UK has substantial difficulties in remedying pollution problems once they have arisen. Thus, in the Maryland solvent recycling plant case even the sanction of closure for a period of months did not prevent the re-emergence of air pollution problems. Similarly, the panoply of regulations in the Florida resources recovery plant case was insufficient to prevent numerous complaints from residents living over half a mile away. Successful prosecution of Leathers Chemicals in the St Helens sulphuric acid works case did not prevent a recurrence of problems. The Yorkshire chemical formulation plant and the Glossop molybdenum smelter provide other instances where the powers of air pollution controllers proved ineffective once pollution problems had arisen.

US air pollution control agencies were more prepared to use sanctions against developers than their British counterparts in the case studies, but the general activity levels of air pollution controllers in enforcement in the two countries were broadly similar. Fines in the US can typically be many thousands of dollars as compared to the usual UK fines of less than a hundred pounds. This is exemplified by the numerous fines levied on the Chevron oil refinery in California and on the solvent recycling plant in Maryland. In Britain, the St Helens sulphuric acid works was the only case where enforcement action was taken in the courts, resulting in a fine of £25.

This difference in penalties would be anticipated where radical new initiatives in control had been taken in one of the countries concerned. However, where penalties are the major operative sanction, the activities of enforcement staff are crucial to the implementation of controls. If air pollution control agency financial and personnel resource budgets are cut as they have been in the United States – and to a lesser extent in the United Kingdom – then enforcement activities will diminish and unpenalised non-compliance will increase. Nevertheless, American resources are by far the most generous. There was, for example, no British equivalent in the cases studied of the two Bay Area Air Quality Management District enforcement staff present almost full-time at the Chevron oil refinery in California.

It appears that the *land use planning agency* has considerable difficulties in implementing controls in both countries. However, when construction takes place where permission has either not been applied for or has been denied it is usually possible to force the dismantling of the offending development through the courts in both countries. This might be expected to be achieved more easily in the UK than in the US because there is less scope for argument about the facts. On the other hand, the greater freedom from central government control may tend to make the land use planning agency more effective in remedying pollution problems in the US than in Britain. Thus, Cecil County succeeded in gaining agreement to relocate in the Maryland solvent recycling case and the City of Midland succeeded in removing the unauthorised Texas asphalt plant whereas Kirklees Borough failed to have the unauthorised tanks removed in the chemical formulation plant case because its arguments were rejected by central government.

The centralised nature of the planning system in Britain, with technical guidance offered to local jurisdictions, has meant that conditions should generally be formulated in a manner capable of implementation and, indeed, there is great emphasis upon this in central government advice (DOE, 1985a). British planning controls over air pollution from new sources should notionally be more effective than American controls in this respect.

Further, land use planning agencies in Britain would be expected to secure better enforcement of planning conditions over air pollution than in the US, because they are better equipped to detect violations, having larger staffs and close relationships with environmental health departments. The complexities of US land use law allow more opportunity for interpretation and the non-enforcement of certain planning violations is almost institutionalised (Haar, 1964). Adherence to planning conditions was secured in several of the British case studies, for example, in the Bolton lead battery works case. However, as a result of a public inquiry, the St Helens authority was frustrated by central government in its attempts to enforce conditions relating to pollution control in the sulphuric acid works case. The difficulties of obtaining satisfactory adherence to the air pollution control conditions set by the state power plant siting section in the Florida resources recovery facility saga is the only US case study illustration of this problem – and here enforcement was left to the air pollution control agencies.

In both countries, it appears that the enforcement of conditions, particularly those relating to the operation rather than the construction of the premises, leaves much to be desired. It is clear that the inadequate effectiveness of both systems is, in some instances, due to the imprecise way in which conditions are couched. Thus, the worst example of ineffective implementation was the inability of the St Helens Council to win an enforcement case under the planning legislation in the sulphuric acid plant case because of the imprecision of the original conditions.

In other instances, however, the personnel involved or the legal redress at their disposal may be inadequate. Conditions relating to physical location or design (eg, number or height of chimneys) are usually capable of enforcement because a single inspection can reveal any problems and there can be little dispute about the interpretation of the permission. There are obviously more British case study examples of this type of control (eg, the Glossop molybdenum smelter and the new town glass fibre works) than American, though the Oregon energy recovery facility was to include a scrubber demanded by Oregon City, the planning agency. Planners in the US and the UK have stated that planning conditions are sometimes no more than a successful bluff that they have little hope of enforcing.[2]

Vigilance of public

The degree of public vigilance does not seem to differ between the two countries. Once sources have been constructed, complaint was vociferous wherever problems arose. It was notable that opposition to most sources arose in both the British and the American cases where construction occurred (Tables 5 and 7). This led to forceful demands to effect enforcement action or to refuse further planning permission for modification in, for example, the Yorkshire chemical formulation plant and the Maryland solvent recovery plant cases. In the St Helens sulphuric acid plant case the election of a member to the local authority to implement improvements was an unusual demonstration of the result of sustained vigilance.

NOTES

1. While both countries have national pressure groups concerned with environmental planning issues, generally the resources of, say, the Sierra Club with its several hundred staff far outweigh those of, say, the Council for the Protection of Rural England. In addition, American pressure groups like the Sierra Club are often able to attract the best graduates of the elite law schools to their Washington offices.
2. Miller and Wood (1983) and numerous interviews by the author with American planners.

9
Conclusions

The first section of this chapter contains an evaluation of the differences between the siting processes for new air pollution sources in the US and the UK. This enables the relative shortcomings of the US and UK systems to be identified in the following section. The final part of the chapter is devoted to recommendations for overcoming these shortcomings in the United States and United Kingdom systems. It is hoped that, together, the conclusions and recommendations will meet the aim of this study expressed in Chapter 1: to assist in improving the utilisation of prospective controls over new or modified sources of air pollution.

EVALUATING US/UK SITING CONTROLS OVER AIR POLLUTION SOURCES

Evaluation of the American and British systems of anticipatory control over new sources of air pollution is as difficult as prediction of the outcome of the siting process. Vogel (1986) summed up his view of the two environmental regulation systems by asserting that neither approach was inherently better. He felt that each stemmed from a particular set of historical circumstances and each had advantages and shortcomings. Earlier, Vogel (1983b) had stated that Britain had managed to strike a balance between the needs of amenity and industry, largely by striving for a consensus, while the United States adversarial rule-making approach left both industry and environmentalists dissatisfied. He asserted that:

> *Each nation regulates the environment in much the same manner that it does everything else.* Compared to that of other capitalist democracies, American public policy does tend to be relatively coercive across a broad range of issue-areas. Similarly, British public policy, when compared to that of other capitalist politics, does tend to emphasise consultation with those interest groups affected by particular governmental decisions (emphasis in original).

Blowers (1984), in comparing air pollution control systems in the US and the UK expressed the view that:

> It is virtually impossible to evaluate the relative merits of the alternative systems of control. The American approach is comprehensive, available for scrutiny, and seeks to avoid inequalities between industries, states or geographical areas. But it is cumbersome, complex and costly. The British system operates flexibly ... but it relies heavily on administrative discretion and a close relationship between government and industry remote from effective public challenge (p 310).

Foley (1963), in discussing the land use planning system in the UK, described Britain as 'corporatist' rather than individualistic and asserted that:

> fierce loyalty to pragmatic approaches and disdain for the theoretical mark British approaches to policy decisions. By showing essential agreement as to goals ... they can concentrate on getting things done to implement these goals ... Americans talk too much about the big 'theoretical alternatives' at the expense of getting on with the job (p 162).

Badaracco (1985) felt that his study of environmental regulation 'vividly displayed the administrative, analytical and political advantages of co-operation over adversarial relationships' (p 161). However, Brickman et al (1985) have attempted to justify the American approach to environmental regulation, while accepting its weaknesses:

> The costs of the American approach to regulation in terms of time, money, and public dissatisfaction are relatively easy to document. The benefits are less tangible. Seen in cross-national perspective, these have less to do with the quality of technical decisions than with much broader attributes of good government (p 313).

What is quite clear is that both the US and UK have, with the exception of the climatically determined pollutants like ozone, broadly similar levels of air pollution. Both countries have made similar strides in reducing pollution levels (Organisation for Economic Co-operation and Development, 1987). However, this has been achieved in the USA at much greater proportionate cost than in the UK, where expenditure on both personnel and equipment is far lower (Chapters 3 and 5). O'Riordan (1979a), in comparing the United States and the United Kingdom, surmised that:

> In the final analysis, it appears that abatement technology will improve regardless of which method is applied, and that the rate of improvement will ... ultimately be political in the sense that the social and economic repercussions of harsh pollution control be set against the gains to public health and amenity.

It would thus appear that the UK system in general is probably more cost-effective than the US system in that similar reductions in pollution levels have been achieved at lower cost. As Asimow (1983) has stated: 'On balance, it appears that the non-adversarial British approach is at least equally effective in improving environmental quality and operates at a lower cost with much less friction'.

Rees (1985) has confirmed these observations:

> Undoubtedly, the British control system has the advantages of being flexible and cheap. Not only are the costs of administration relatively minor compared with those in the United States, but also there is little expenditure on court cases. In addition, the economic costs to industry of meeting the regulations appear to be much lower, with little difference in the outcomes *vis-à-vis* the reduction of pollution (p 371).

The system of controls over new sources must, to a large extent, be typical of the general air pollution control systems in the two countries. However, if the British system of controls over new sources administered by air pollution control agencies is more *cost-effective* than the American, it does not necessarily follow that the whole siting process is more efficient, equitable or effective.

Efficiency

The degree of efficiency achieved in the permitting process depends on several factors. One is the cost of the transaction to the developer. In general, efficiency demands a minimisation of the number and duration of permit procedures and, in this sense at least, the British system must be more efficient than the American system. As Brickman et al (1985) have commented:

> With respect to the administrative costs of complying with regulations... there is little question that US industry has to spend considerably greater sums than its European counterparts (p 312).

Thus, the UK planning system involves a single permit. This is normally granted within two or three months; though more complex applications, such as those involving significant air pollution, often take longer. Although much of the cost of land use planning is borne by ratepayers and taxpayers in the UK, the developer pays a fee and may have to bear substantial costs in providing information and, possibly, in suffering delay. These costs and delays can become very significant if lengthy public inquiries are held but application processing costs are usually a tiny proportion of total project costs.

In the United States, though the land use control system costs the taxpayer very little, the costs to the developer of obtaining a number of land use permits can be very substantial. More land use permits may be required and more information has to be provided than in the UK, increasing the developer's costs substantially. For example, the $200,000 paid by Chevron for its environmental impact report on the California oil refinery modification was more than any UK developer had to pay. The length of each land use planning agency's decision process – a determinant of the developer's costs –

in the United States case studies was broadly similar to that in the United Kingdom.

The same general picture can be drawn of the transaction costs associated with the air pollution control agency. In general, the fees to be paid for applications to various agencies and the costs of providing voluminous information, of fighting appeals and court cases, of delays, and of generally negotiating a pathway through a maze of environmental regulations are all likely to be much higher in the United States than in the United Kingdom. Thus, the cost to BECO of its attempt to site an oil refinery in North Carolina was put at $3,000,000 whereas Metro and Wheelabrator-Frye together spent over $2,000,000 in the Oregon energy recovery facility case.

Another factor determining efficiency is the cost to the developer of complying with all the relevant new source environmental regulations. However, while technical control, monitoring and administrative costs are higher in the United States, efficiency also requires that the costs of pollution damage from the proposed development be minimised and borne by the developer. It is therefore possible that the more comprehensive and stringent mitigation conditions often imposed in the United States may be more efficient than British conditions because the damage likely to arise from new sources of air pollution should be correspondingly less. Thus, the developers of the Oregon energy recovery facility claimed that the very high level of pollution control agreed would minimise damage costs borne by third parties.

The granting of permits by both US and UK air pollution control agencies, with virtually no refusals, cannot always lead to an efficient outcome since there will be circumstances in which the costs of damage arising in a particular location will be greater than the benefits of the new source (the conditions inappropriate case in Figure 2). Similarly, the general lack, in both countries, of substantial variation of conditions to reflect location must be economically inefficient because the capacities of different environments to assimilate a given level of pollution vary. Because the intensity of land use tends to be lower in the US, the cost of damage from similar pollution sources should be less than in the UK.

In some instances the outcome may be inefficient in that the developer has to pay too high a price (the conditions may be too strong – in the terms used in Figure 2). The greater degree of public participation in the United States may sometimes decrease efficiency by resulting in the tightening of conditions already stringent enough to achieve the optimal level of pollution. There must be a suspicion that the conditions demanded of BECO in the North Carolina oil refinery case, and the guarantee of an inflation-adjusted sum of $70,000 per annum to enable collection of garden refuse extracted from Metro in the Oregon energy recovery facility case, were too strong and likely to lead to an inefficient outcome. Such outcomes may not be unique to the US because it might also be argued that Chloride utilised controls which were unnecessarily stringent in the Bolton lead acid battery case.

On the other hand, it seems that decisions may sometimes be inefficient in the UK for a different reason. Insufficient control may be applied by the air pollution control agency (ie, the conditions may be 'too weak') and the residual pollution problem may result in high damage costs. This certainly appears to have been true, for example, in the St Helens sulphuric acid works case where very severe pollution damage occurred. In other instances, such as the Yorkshire chemical formulation plant case, no anticipatory controls were used and subsequent pollution damage costs showed that this was an inefficient solution.

One inefficient outcome of a land use planning agency's decision is the refusal of a permit to a development which would have caused few environmental problems, perhaps because of the political nature of a pollutant. This is more likely to happen in the United States than in the United Kingdom and the Louisiana creosote storage plant and the Texas asphalt plant could be quoted as examples of this type of decision. Another sub-optimal outcome is the situation in which permission is granted by the land use control agency to construct a source which is fundamentally incompatible with its neighbours and incapable of being ameliorated by conditions. The case studies indicate that this is probably more likely to arise in the United Kingdom, an example being the original decision in the Glossop molybdenum smelter case.

It seems that land use planning agencies are more likely to consider air pollution in the UK than in the US, if only because the environmental health officers of their own local government will press them to do so. The existence in the UK of central government guidance on conditions and of its planning inspectorate ensure that professional planners are very aware of government policy when determining conditions and this leads to considerable uniformity. Therefore, land use conditions are perhaps more likely to be too weak rather than too strong (Figure 2), whereas the reverse is probably true in the USA. On the whole, the types of conditions negotiated by the City of Richmond in the California oil refinery modification case and by Oregon City in the energy recovery facility case are probably more efficient than many British outcomes, since some compensation is being paid to the local community for the pollution and other costs incurred.

Generally, whereas higher transaction and control costs will increase the developer's expenditure in the USA, and may thus be inefficient, the costs of damage in the UK may be higher and also be inefficient. Because anticipatory controls are relatively cheap to implement, this study indicates that, so far as air pollution from new sources is concerned, the American system may be marginally more efficient than the British.

Equity

Both outcome and procedural equity demand that a full array of environmental protection measures is brought to bear and that as much relevant information as possible is made available to both the control agencies and the public before binding decisions are reached. Only by full disclosure, meaningful public participation and the fair operation of agency processes can equity be achieved and, as importantly, be seen to be achieved. It is therefore probable that equity is higher in the United States than in the United Kingdom, since it is more likely that, in the US system, mitigation measures, compensatory actions, etc, will be negotiated to ensure that the few do not suffer without compensation for the benefit of the many when a new stationary source of air pollution is contemplated.

The *outcome equity* of the air pollution control agency's decision is perhaps likely to be higher in the US than in the UK. In general, the developer is equitably treated in both countries by being involved in negotiations determining the eventual outcome and hence influencing it, by being granted a permit and, in relation to other developers, by not being given a significantly location-dependent decision. The local residents, on the other hand, are inequitably treated by both the virtually certain grant of the permit and the relative uniformity of conditions imposed in both countries irrespective of the characteristics of the specific location. Because of rather stricter conditions on new sources in the US (and lower population densities) local residents are likely to be less subject to pollution from a new stationary source of air pollution than in the UK and hence to be more equitably treated.

Again, the equity of the land use planning agency's decision to locate a new stationary source of air pollution may be higher in the United States than in the United Kingdom because the conditions imposed on the development may be more stringent. The greater availability of benefits to local residents in the United States because of the smaller size of local authorities and the nature of the local revenue base leads to a more equitable outcome in that country. The residents in the immediate locality of a new source of air pollution in the UK, however, may be inequitably treated because they suffer damage while they gain little or nothing by way of compensation.

On the other hand, it is probable that the developer in the United States is more likely to be treated inequitably than in Britain by having a development refused, on political or emotive grounds, which might legitimately have been permitted. The larger population of most local jurisdictions in the United Kingdom will make it more difficult than in the USA for the minority directly affected by pollution from a new source to convince the majority to reject it, especially where the benefits to the larger population are substantial. An existing company offering substantial local employment will seldom lose a battle to expand or modify its premises, in either the USA or the UK, irrespective of the air pollution resulting.

The *procedural equity* of the air pollution control agency's decision is also likely to be higher in the United States. There are far greater opportunities for participation and for objection in the US than in the UK, and the outcome can be, and is, influenced by the representations of third parties. The activities of

Citizens for a Better Environment in the California oil refinery modification case in demanding an appeal and of the local residents in the Texas asphalt plant case in requiring a permit exemption hearing have no equivalent in the UK. There are, of course, participation costs but these can yield substantial dividends in the stringency of conditions applied to air pollution permits.

Similarly, the procedural equity of the land use control agency's decision is likely to be higher in the United States before any appeal because of the greater opportunities for public participation and objection in that country and the right of third party appeal against decisions. In the USA, it is often possible to appeal to the courts against decisions already subjected to appeal to the local government. Thus, the legal appeals by the developer in the Maryland solvent recycling case and by objectors in the California chemical works case had no equivalent in the British cases. The appeal to the Oregon Land Use Board of Appeals by the objectors, however, was not dissimilar to a British planning appeal to central government.

Only if the government refuses a development (as in the Bolton incinerator case) and the developer appeals or if the development is so obviously controversial that it is called in by central government (as in the Cheshire fertiliser plant extension case) do objectors have the formal right to be heard in land use planning decisions in Britain: though they can and do make written, and sometimes oral, representations before a local authority has made its decision.

If an appeal is heard in the UK, the developer, the land use planning agency and the objectors all have ample opportunity to present their cases and cross-examine the evidence of others, whereas their rights are more limited in the United States. Major British inquiries bear little relationship to American hearings on zoning or land use permit decisions. Rather they are comparable to, but go beyond, American rule-making procedures. They are prolonged, highly contentious, and governed by strict rules of procedure. Though none of the four public inquiries in the case studies was lengthy, they may last many months in particularly controversial cases. Because time and expertise are required to participate in inquiry proceedings, public involvement is often more apparent than real (Pearce et al, 1979). However, this is generally preferable to the sometimes farcical time-limited involvement of the American citizenry in public hearings, which may be limited to a five minute presentation.

Procedural equity is obviously served by the objectors' possession of adequate resources and the opportunity to utilise them in disputing the granting of permits. The avenues open to the objectors in the Dow chemical plant case or to the opponents of the energy recovery facility in Oregon were much more extensive than those available, for example, to the objectors in the Cheshire fertiliser plant extension case. Notwithstanding what therefore appears, on balance, to be a superiority in procedural equity in the US compared with the

UK, Vogel (1986) reported that public perception of procedural equity is greater in the United Kingdom:

> Siting decisions are more likely to be accepted as legitimate in Britain – even when they go against a particular developer – because they are perceived as having been decided on their merits. In America, on the other hand, the substantive issues often tend to be overshadowed by procedural ones (p 189).

The same is probably true of the perception of objectors.

Effectiveness

Blowers (1984) has discussed the problems of evaluating the effectiveness of the US air pollution control system:

> On the one hand it is argued that it has secured cleaner air and protected clean areas from worsening conditions. On the other it is argued that it has slowed economic growth, precluded the development of more effective control technologies, and focused on new sources while ensuring the maintenance of existing high pollution sources (p 308).

These evaluation problems apply equally to the United Kingdom system in general and to controls over new sources in both countries in particular.

It would be expected that a system in which controls were discussed co-operatively should lead to more effective implementation than one in which they are imposed by rule. Thus, Lundquist (1980) contended that:

> 1 The more open and conflict-oriented the political system the more immediate and substantial the response to problems of environmental quality but the less substantial and successful the implementation of adopted policy alternatives.
> 2 The more closed and consensus-oriented the political system, the slower and more incremental the response to problems of environmental quality and the more deliberate and successful the implementation of adopted policy alternatives (p 34).

He concluded that there was, in fact, little discernible difference in effectiveness between the USA and Sweden, a conclusion which would hold were the UK substituted for Sweden as the example of the more closed system. He felt that American implementation had been better than might have been anticipated, largely because of the constraints and incentives provided by the Clean Air Act itself and in particular because of the numerous provisions for public participation and citizen litigation afforded by the act.

Lundquist's view of the lack of substantial difference in the effectiveness of environmental regulation between countries with open and relatively closed implementation procedures was confirmed by Knoepfel and Weidner (1983). They further argued that there is likely to be a higher degree of compliance with formal conditions in closed systems, where bargaining by consent, rather than coercion, is more important.

The effectiveness of the outcome of an application to construct a new air pollution source is likely to be much greater if the mitigation measures are agreed during negotiations, rather than imposed unilaterally, since implementation inevitably relies on a measure of voluntary co-operation. Similarly, conditions requiring the installation of control equipment may prove more effective than those involving working practices. Again, the more different agencies that are involved, the greater the chance that implementation will not be complete and, thus, that effectiveness will be compromised. These factors would tend to suggest that the effectiveness of the controls imposed on air pollution from British new sources should perhaps, be greater than those on their American equivalents.

On the other hand, greater public involvement in the US may lead to closer scrutiny of the potential pollution problems and the control measures agreed and hence to a more effective decision. Though the plants were never built, the activities of objectors acting to secure requirements for the activated carbon filter in the Maryland solvent recycling plant case and for the sulphur dioxide scrubber in the Oregon energy recovery facility case exemplified this. The greater array and stringency of both anticipatory and retrospective controls and penalties available in the US should also result in better effectiveness than in the UK.

It is worth noting that, if the proposed development would have resulted in large damage costs and is stopped by objectors, their action will have been an effective anticipatory pollution control measure. This outcome is perhaps more likely in the United States than in Britain and examples include the Childs solvent recycling plant in Maryland.

In fact, the evidence from this study suggests that the implementation of controls imposed on new sources of air pollution is not markedly more effective in one country than the other. The eventual reductions in pollution damage achieved in most of the cases examined where construction occurred indicate that both systems can be effective, though frequently only after a delay. This study also shows that pollution abatement perhaps depends more on the attitude of the management of the organisation responsible for the pollution source than on any other factor in both countries.

IDENTIFYING RELATIVE SHORTCOMINGS IN THE US AND UK PREVENTIVE CONTROL SYSTEMS

It is apparent that there are advantages and disadvantages in the systems of control adopted in both countries. It follows that it should be possible to improve both by comparing one with the other. Asimow (1983) believed that:

Each country has something to teach the other. American rulemaking procedures have improved the quality of rules and furnished a sense of participation very satisfying to the persons who must live with the rules. Similarly, British inquiry procedures, for all their defects, have brought the people closer to government decisions having critical effects on their lives. Neither country will, or should, abandon these procedures, though they must be pruned from time to time.... Both countries should begin the process of judicious sampling of the other's fumbling attempts to involve the public in critical administrative decisions.

Vogel (1983a) concurred:

The British would probably benefit from more clearly defined standards, a greater willingness on the part of government to prosecute, and increased opportunities for public participation in the policy process. For America, on the other hand, more flexible regulations and closer consultation with business might well result both in more sensible rules and more reasonable enforcement policies.

In comparison with the United Kingdom the *developer*'s transaction costs can be seen to be too high in the United States. The greater recourse to the courts in the United States further increases the costs of siting a new air pollution source. It would be argued further that the position of the US developer in the siting process is inequitably weak in relation to that of potential objectors in comparison with his UK counterpart. In the United Kingdom, the larger size of local authorities and the system of local government finance means that the developer has inadequate opportunity to compensate those directly affected by pollution.

The *air pollution control agency* in the United States has to administer an overly complex and expensive system of controls which can be somewhat arbitrary in nature, with relatively little scope for discretion, in comparison to the UK. For example, it does not extend to permitting major sources to locate in non-attainment areas, which are often so classified on the basis of inadequate monitoring information, without the provision of offsets – the calculation of which is often nebulous. However, there is usually too little consideration of the immediate effects of pollution on near neighbours of the proposed source compared to the situation in Britain. The agencies involved have to contend with rules that change too frequently and with too high a staff turnover, in comparison with their less numerous British counterparts.

In the United Kingdom, on the other hand, air pollution control agencies take decisions with inadequate regard for ambient pollution levels. While giving consideration to the proximity of sensitive receptors, the control response tends to be limited to requiring higher chimneys to promote greater dispersal. HMIP possesses comparatively comprehensive anticipatory powers but those of local authority environmental health officers are so inadequate that they frequently request their colleagues in planning departments to incorporate air pollution control conditions in planning permissions or agreements. The process of negotiation with the developer is carried on with too little of the countervailing public input which has proved so valuable in moderating air pollution control decisions in the United States. In comparison with the United States, where some third party appeal rights exist, there is insufficient public right to appeal against air pollution control decisions.

The American *land use planning agency* is typically too small, too parochial and too politically influenced in its operation in comparison with its British counterpart. The controls at its disposal are fragmentary and weak and are usually deployed with insufficient regard for the provisions of a land use policy. Too many land use decisions are taken by appointed representatives without the benefit of recommendations by professional planners. Staff quality is often poor in the land use planning agencies and turnover is too high because of the fee-funded and federal programme grant-aided nature of much of the work. There are too many conflicts between the various, and overly numerous, state and local land use planning bodies.

The United Kingdom land use planning system has features that lead to too strong a bias towards decisions in favour of the developer in comparison with that of the United States. The public has too little access to information in advance of most UK planning decisions. Its comparative lack of access to environmental impact data has been particularly noteworthy. While public inquiries can be lengthy and expensive for those involved, the contrast between information availability in cases where such inquiries are held and that obtaining at the time when initial decisions are made by the local planning authority is often too great. Compared with the US, there is inadequate third party right of appeal against the decision of the land use planning agency.

The taxation system in the UK, which permits charities to increase the size of annual gifts they receive by reclaiming tax, is insufficiently favourable to environmental pressure groups and these groups are generally inadequately funded and staffed when compared with their more numerous United States counterparts. *Objectors* are consequently frequently at a relative disadvantage in the UK.

The *authorisation process* in the United States, in comparison with the United Kingdom, is thus too complex, too legalistic, too expensive, and subject to too many variations and changes in rules. In Britain, on the other hand, the authorisation process is too secretive and furnishes too few anticipatory controls in comparison with the United States. The system is, with the exception of the public planning inquiry, too biased against objectors and provides too few opportunities for compensating those suffering the pollution from new sources.

The *implementation of* planning and air pollution *controls* is probably too weak in both the US and the UK, partly as a consequence of budget cuts, though it seems likely that there is too low a degree of voluntary compliance by the developer with the conditions imposed by both types of control agency in the US. However, the availability of retrospective powers, the incidence of formal enforcement action and the level of penalties imposed are all too low in the United Kingdom when compared with the United States. Because the numbers of staff allocated to the enforcement of air pollution conditions are too small in the UK, there is a too great a reliance on public complaint as an indicator of problems and insufficient monitoring.

IMPROVING THE PLANNING OF POLLUTION PREVENTION

Hall (1982) has offered general advice about the assessment of new air pollution sources and other important projects. He believed that, having decided whether a new project is viable in financial terms, the positive and negative effects of the investment on other people should be considered. Despite the difficulties, the groups upon which the costs and benefits would fall should be identified, to judge the distributional consequences. These external costs and benefits should then be allowed to affect the decision as to whether to proceed or not. His analysis led to the recommendation that: 'Above all, because of the great uncertainty inherent in nearly every planning decision, the golden rule remains: do the minimum necessary, and leave tomorrow's decision for tomorrow' (p xxvi).

A corollary of this is the necessity to involve potential objectors in the siting process. As Knödgen (1983) has reported, there are significant advantages to be gained from involving the public:

> Firms and state or local officials who attempt to overlook public concern, or to hinder or reduce their sometimes time consuming participation even risk provoking increasing environmental conflicts. A more efficient strategy for locating successfully seems to be careful consideration of environmental protection in selecting the appropriate site and the improvement of public participation.

Much has been made, particularly in the USA, of the potential role, in certain circumstances, of environmental mediation in reaching a solution to siting disputes. Mediation involves the assistance of a mediator in negotiations between the parties in a dispute over a new development and requires a willingness to compromise and utilise environmental mitigation. While it is not easy to state precisely when and if mediation will help negotiations towards completion, there appear to be four prerequisites to its success: a stalemate, or the recognition that stalemate is inevitable; voluntary participation; some room for flexibility; and a means of implementing agreements (Talbot, 1983; Bingham, 1986). These prerequisites appear to apply in only a small minority of siting decisions.

Similarly, financial compensation or remuneration in kind may also have a role to play in both the US and the UK. Frieden (1979) has described a case where monetary compensation was paid in California and Knödgen (1983) one in West Germany. Though such payments may be controversial, and though there may be considerable problems of determining who should receive the payments, the use of compensatory measures can increase economic efficiency, as pointed out in Chapter 1.

One of the more interesting findings of this comparative study is the revelation of how rarely planning techniques for the reduction of air pollution were used in either the US or the UK. While some of the available techniques were inappropriate in the case study circumstances, there is clearly a serious lack of knowledge by planners about the available control techniques. There is real scope for the development of planning guidelines on the use of pollution control techniques such as buffer zones, the design and arrangement of buildings, siting with respect to terrain, etc. Equally, there is a need for compilation and collation of the scattered existing knowledge on planning techniques for controlling pollution. The resulting information might be presented as a manual for practising planners which would be of equal utility in the United States and the United Kingdom.

Equally, in both countries there is a need for a source-book for planners on air pollution information, on consultation opportunities and requirements, and on other sources of advice. This publication would be specific to the country concerned[1]. Such source-books, together with the techniques manual, would do much to equip land use planning agencies in both countries better to undertake their responsibilities in new source control.

Perhaps the greatest impediments to better performance in authorising and then supervising new sources of air pollution in both countries are attitudinal – whether attitudes of intransigence or indifference. The attitude of the developer in the United States has been characterised as sometimes lacking in concern for the environment. The same could be said of some developers in the UK. There is a need for the introduction of environmental issues involving industry, and particularly the new source authorisation system, into business school curricula in both countries. Similarly, air pollution controllers need education and training in the role of land use planning agencies in air pollution control and, of course, planners should have equivalent knowledge about the role of air pollution control agencies as well as extended training in the use of appropriate planning techniques. The attitudes of politicians taking land use decisions can also be very important and could be improved by short courses and other such methods.

There is scope for considerable further research in relation both to understanding the effects of the anticipatory control systems on the siting process in both countries and to improving their functioning. There is obviously considerably more work to be undertaken in relation to the six indicators about the nature of the authorisation process. The undertaking and analysis of brief case studies of the siting of air pollution sources in the United States and the

United Kingdom would be necessary to isolate further the roles of the various factors and their relative weights in order better to predict outcomes.

There is a need for further investigation of the costs of air pollution control and of air pollution damage, especially in the UK. Similarly, there is scope for further interdisciplinary research on the various costs and benefits of planning decisions involving new stationary sources of air pollution and on the reconciliation of environmental and economic objectives by the use of mediation, compensation and planning gain approaches.

United States

Duerkson (1983) made several recommendations for the 'quiet reform' of the siting process for both new air pollution sources and for industry generally in the United States. He suggested that federal, state and local agencies could improve management by such methods as the designation of lead agencies, the appointment of a single project manager, the use of agreed timetables for decision making, etc. He felt that companies needed to assess potential environmental impacts earlier, to use a more open project-planning process with early government and citizen participation, and to increase the use of alternative methods for settling siting disputes such as mediation and mitigation. Specifically, he thought there was a need for greater use of scoping (Chapter 3), the compilation of permit guides and earlier release of information. Duerkson concluded that:

> The most promising methods of improving the efficiency of the environmental and land-use regulatory system are those that stress co-operation and negotiation. These innovative initiatives ... offer relief without deforming the system, but they will work only if *all* the players in the siting game co-operate (p xxviii, emphasis in original).

Bardach and Kagan (1982) recommended working within the current system to make it more reasonable, with inspectors having the discretion to fulfil the spirit rather than the letter of the law, not unlike the British procedure. Seley (1983) presented several new approaches to public facility planning: for example, dispersion and coordination to ensure that certain areas were not overburdened with unpopular facilities. However, he stressed that improving the current methods of siting: muddling through; conflict resolution; compensation; and incentives, was necessary (pp 177–191).

O'Hare et al (1983) were more pessimistic than Duerkson and others about conventional reforms but argued for many of the same methods: negotiation, earlier provision of information, mitigation, compensation and mediation. They proposed that a siting agreement be signed between the developer and the local community before formal permits were applied for:

> This agreement specifies the nature of the project in detail, with local concerns taken account of, and the operating rules that will govern the community and developer both. It is established either by the parties themselves or, if negotiations run aground, by the arbitration panel (p 180).

These concerns would include, but not be confined to, air pollution and other environmental problems.

Morell and Magorian (1982) insisted that the siting decision must be left in local hands but argued that extensive mitigation of impacts, negotiation on non-environmental concerns and compensation could be successful in winning local approval; though each technique had its limitations and some opponents would remain unconvinced whatever concessions were made by the developer. Agreement could:

> ... best be accomplished by extensive mitigation of their adverse impact accompanied by negotiated compensation. A siting policy that aims to reduce inevitable losses to an acceptable level for those who are harmed and to build citizen respect through balanced sequential, and timely procedures is not only carefully conceived but, more importantly, should also prove effective in clearing the political impediments to successful siting in the United States. This structure is in accord with the political realities of ... siting controversies, and very different from the myth of preemption (pp 188–189).

This study confirms that a number of these suggestions for reform would undoubtedly improve the process of authorisation of new sources of air pollution. This is particularly true of those relating to the use of lead agencies, the earlier release of information by companies, the increased use of mitigation measures – and perhaps of mediation – and the use of compensation payments to those directly affected by air pollution and other adverse impacts. Comparison with Britain also indicates that greater co-operation in implementing controls could prove fruitful. In particular, the building of consensus between the *developer* and all the other parties involved by the use of state co-ordinators, and external mediators in appropriate circumstances, could expedite many siting disputes, reduce developers' transaction costs and reduce the incidence of resort to the courts.

A number of suggestions for more radical reform can be advanced. The most obvious of these, following comparison with Britain, is the desirability of local government reform. The structure of the local government system in the USA is not dissimilar from that of the UK in the mid and late nineteenth century. The situation in which the mean population size of counties and municipalities is only just over 10,000, with much fragmentation of local jurisdiction within continuous urban areas, is obviously ripe for reform. Rationalisation would also need to take account of the extra and ever increasing number of townships, school districts, and other special purpose districts – presently over some 60,000 in total (Chapter 3). Hagman (1975) has advocated a thorough-going reorganisation of American local government, involving much larger counties and constituent cities of a size up to 250,000 persons. He believed that 'massive reorganisation of local governments is the priority land use planning need in America' (p 123). While not agreeing that the British reorganisation of 1974 achieved all the objectives he put forward, he did regard it as a model of what could be done, given the will. Such reform would reduce parochialism and local taxation anomalies while providing a firmer base for the recruitment and retention of adequate professionally trained staff.

This study shows that there is a clear need to simplify the system operated by the *air pollution control agency* for new sources. This must be achieved without weakening its performance capability. One area of improvement would be to eliminate the use of modelling exercises based on notional rather than monitored ambient air pollution levels. This approach has been seen, as in the Oregon energy recovery plant case, to lead to enormously complex and almost incomprehensible prevention of significant deterioration calculations. There must also be a very strong argument for combining the new source performance standards, best available control technology and lowest achievable emission rate requirements into a single set of standards for major and minor sources, while retaining the offset requirements currently applied in non-attainment areas (Chapter 3).

Similarly, the division of permit responsibilities between federal, state and, sometimes, local agencies leads to needless duplication. Consideration should be given to allocating all clean air permit functions to the states (which might then delegate them for particular areas) subject to active monitoring and the right of intervention by the Environmental Protection Agency. Such an allocation would reserve to EPA certain sanctions for inadequate performance such as grant withdrawal and construction moratoria.

Again, by comparison with the UK, the performance of the United States anticipatory control system would benefit from a strengthening of the system operated by the *land use planning agency*. At a general level, the case studies indicated that the opportunity exists to improve the systems as between states, with Louisiana, Texas and North Carolina appearing to have weaker controls than Oregon and California, for example. Specifically, wider use of zoning systems and of requirements that zoning be in accord with the land use plan would remove many difficulties over the siting of new air pollution sources. The case studies also show that there is often a need to increase the competence and staffing levels of local land use planning agencies. The experience of those communities affected by proposals to construct new sources of air pollution and by the environmental impact of other major industrial proposals could provide the focus for a national campaign by environmental pressure groups for improvement of the land use planning system. Local hearing procedures need to be improved, perhaps by agreeing who should be called in advance and extending the length of time allowed for submissions.

There is clearly also scope for the simplification of state land use permit procedures. There would be particular benefit in reducing procedural delay by the operation of the large number of state permit procedures co-ordinated by means of such things as joint hearings. The increased use of environmental impact assessment as a shared basis for evaluating the environmental effects of a new source could be one method of facilitating this joint operation of permit systems. Notwithstanding the comments of Morell and Magorian (1982) quoted above, there may well need to be provision to allow state views to override those of local governments in certain circumstances. This could perhaps be accomplished by the setting up of a state appeal system (equivalent to the British central government planning appeal), to which the developer could have recourse as of right and which *objectors* could harness if a sufficient proportion of those resident within a certain radius of the proposed source petitioned for it.

Clearly, many of these proposals to improve the *authorisation process* in the US are likely to prove difficult to implement. This is why the 'quiet reforms' are so attractive, though this study indicates that the current use of one-stop or co-ordinated state permit procedures is not proving very effective. The increased use of mitigation and compensation measures should improve the climate of opinion for the achievement of the various reforms suggested and help to shift the balance of advantage in siting some new sources of air pollution towards the developer.

There is clearly a case for improving the *implementation of controls*. Greater priority needs to be given by the state air pollution control agencies to the enforcement of new source conditions by the allocation of more staff and greater use of sanctions where co-operative bargaining proves ineffective. The enforcement of both local and state land use permit conditions could with benefit be strengthened by increasing the range of implementation powers, the level of penalties and the priority of ensuring compliance in land use agencies.

There is a need for considerable research on appropriate formulations of local government reform, on the technical implications of simplifying the current air pollution control law and on the feasibility and implications of delegation of air pollution control responsibilities. Similarly, there is scope for substantial investigation of the repercussions and acceptability of state override of local land use planning decisions, of strengthening planning controls and of simplifying state land use permit procedures.

United Kingdom

Several ways in which anticipatory control over air pollution in the UK might be improved and siting problems eased have been suggested. The Royal Commission on Environmental Pollution, as well as urging the adoption of ambient air pollution guide values in the UK and the formation of 'Her

Majesty's Pollution Inspectorate', dealing with all types of complex pollution problems (RCEP, 1976), suggested that, as the British rating (local tax) system offers inadequate relief to residents suffering pollution burdens, compensation for the effects of environmental damage should be utilised in Britain:

> We recommend that the Government should examine the appropriateness and feasibility of adapting UK law and administrative procedures to provide for some form of discretionary compensation or inducement to individuals or communities affected by . . . sites, such that the costs fall ultimately on those who generate the waste (RCEP, 1985:134).

This recommendation has become more urgent with the replacement of rates by a community (per capita) charge.

Wood (1976) suggested improving planning control over pollution by increased provision of information, the use of clearer objectives and standards, increased powers under planning legislation, more effective consultation and more training to increase awareness about potential pollution problems amongst those involved in planning decisions. Miller and Wood (1983) argued that a change of attitudes towards pollution was required and suggested that the developer should be obliged to demonstrate the need for the proposed pollution source. They recommended the use of standards, using the analogy of the recommendations of the International Commission on Radiological Protection, which had been accepted by the UK Government:

> Central government has therefore endorsed a system of regulating radioactive pollution which includes quantitative standards and demonstrations of 'net benefit', yet equivalent controls have not been readily accepted in the case of inactive pollutants (p 226).

The latest change in local government in Britain is the abolition of the metropolitan county councils, a measure which is likely to run counter to the best interests of environmental management (Wood and Jenkins, 1985). The reintroduction of some form of sub-regional local government in the metropolitan areas may well prove necessary to reduce air pollution problems in conurbations.

The payment of compensation by the *developer* to those directly affected by air pollution from a new major source, perhaps from permit fees charged by air pollution control agencies and/or by land use planning authorities, might reduce conflicts in certain cases and would be a desirable extension of the UK system. The likely damage could perhaps be assessed by the air pollution control agencies on the basis of proximity to the source and the nature of the emissions.

There is scope to improve the system operated by the *air pollution control agency* and some reforms are afoot. While it is obvious that the UK should avoid the complexity of the US system, the use of standards needs to be increased. This might be done by tightening the current European standards and by extending the coverage of ambient limits to pollutants such as ozone. The corollary of requiring the use of offsets or prohibitions on new source development in areas exceeding the standards merits careful and serious consideration. There are, of course, numerous difficulties to be overcome in applying a system of stringent mandatory standards but the potential advantages are too great to be dismissed by clamorous references to the flexibility of the British system and by resort to taller chimneys. The need for greater investment in pollution control equipment will have to be faced (Weidner, 1987). The anticipatory control powers of the environmental health officers of local authorities are woefully inadequate by comparison with those of HMIP and there is a pressing need to implement the proposals to grant local authorities equivalent powers over non-registered sources (DOE, 1986).

The UK public's right to information ought to be increased to counteract the inevitable tendency of air pollution controls to be influenced by the relatively uncontested representations of developers. Such public information needs to be made available prior to the grant of permit, needs to be more comprehensive than the current proposals indicate (DOE, 1986), and should preferably be supplemented by a third party right of appeal against the air pollution control agency's decision in certain circumstances. These might arise when a given proportion of residents living within a specified distance of the proposed plant demanded such an appeal.

HMIP is a welcome development, as its scope has been broadened to cover major pollutants affecting all media. In parallel, the proposed changes to air pollution control should be extended to the delegation of many of the Inspectorate's functions to local authorities to avoid the duplication of effort, and conflict of objectives, that so often occurs at present. HMIP would obviously need to retain oversight responsibilities and intervention rights – similar to those of the US Environmental Protection Agency – because the competence of local authorities would vary considerably, especially in the initial stages of such a modified system. The undoubted cost-effectiveness of its predecessor should not prejudice the need for adequate manpower. Analogy with American experience suggests that there may be a case for extending the HMIP concept to an EPA-style government department responsible solely for the environment (ie, without conflicting responsibilities) with a Cabinet minister.

Finally, because many of these reforms would be costly, the proposals to introduce a system of permit charges, as in the United States, to contribute to the costs of air pollution control, are to be welcomed (DOE, 1986). The precedent for construction charges has been set in the UK with planning application charges: the case for air pollution permit payments is much greater, as it can be justified by the polluter pays principle. The fees ought to be sufficient to pay the compensation charges mentioned above, together with a proportion of the air pollution control agency's expenditure.

It is quite obvious from American experience that, even if air pollution control agencies gain comprehensive anticipatory powers over air pollution, the *land use planning agency* will be left to make the decision as to whether to permit the development or not. There is a need for changes in the land use planning system though it should be less necessary to append air pollution control conditions to planning permissions. The requirement would remain, however, for much greater provision of information to the public at the early stages of an application to locate a new source or to extend an existing source of air pollution. Similarly, public access to planning files and not just to committee reports and a right for interested members of the public to be heard at planning committee meetings, perhaps on the basis of a petitioned request, would be valuable in improving procedural equity[2]. A parallel right of public appeal against planning decisions should be considered, perhaps on the basis of a request from a specified proportion of residents living within a certain distance of the proposed site. The agreement, in advance, of how long each participant at planning inquiries should have for both the presentation of his case and for the cross-examination of opposing witnesses might also prove a useful reform.

There is a strong case for amending the General Development Order[3] to make consultation by the planning authority with the air pollution control agencies mandatory when a proposed new or extended source is under consideration. Consultation over major pollution-sensitive developments in the proximity of existing major pollution sources should similarly become mandatory. There is also a need to tighten the order to prevent the intensification of use of premises, without the need for planning permission, which can lead to increased air pollution: as demonstrated by the case study concerning the Yorkshire chemical formulation plant. The same case study shows the need for an alteration to the Use Classes Order [4] to prevent change from an environmentally innocuous use of buildings or land to one which pollutes neighbouring areas, while avoiding a requirement for planning permission.

Wider use of the valuable anticipatory planning tool of environmental impact assessment is to be welcomed. While it has been possible for a planning authority to ask a developer for a comprehensive statement of the pollution implications of a proposed development, the onus of preparing an EIA report falls on the developer. Formal EIA should ensure wider provision of information, support and facilitate increased consultation, and ensure that environmental impacts are considered in detail in planning applications involving air pollution sources. The implementation arrangements require HMIP to be consulted where air pollution impacts are likely (DOE, 1988). It is to be hoped that HMIP will adopt a positive view of this useful requirement.

The most important reform to help redress the balance towards *objectors* would be the passing of a comprehensive freedom of information act. United States experience indicates that this could give pressure groups the opportunity to participate in a more meaningful fashion in the siting of new sources of air pollution, notwithstanding recent and proposed improvements. This could ensure fuller consideration of the effects of air pollution and hence better anticipatory control. There is also a case for encouraging environmental pressure groups to focus on anticipatory controls over pollution sources by means of grants.

Even with the introduction of these suggested reforms, economic factors may still tend to outweigh environmental factors in *the authorisation process* but the ramifications of decisions should become more explicit. It could be expected that local authorities would be influenced to greater use of planning conditions or agreements increasingly aligned towards redressing this balance and to overcome these disbenefits. This study indicates that such reforms are overdue and that deregulation of, and budget cuts in, the implementation of anticipatory controls over new air pollution sources are likely to prove counter-productive (Wood and Hooper, forthcoming).

There is, as in the USA, a need to improve the *implementation of controls* over new sources of air pollution. It is clearly desirable to increase the scope and utilisation of air pollution control agency enforcement powers and the amounts of penalties for non-compliance. This would help to ensure that developers maintain the priority given to pollution control once their permits have been received. Similarly, the range and the strength of local planning authority enforcement powers needs to be extended. The priority awarded to enforcement within planning departments also needs to be raised.

These recommendations in regard to the UK system raise matters which provide scope for a substantial research effort. In particular, the reconciliation of meaningful air quality standards with the best practicable means concept of air pollution control and its extension to best practicable environmental option require analysis. The ramifications of using standards which may prevent new development in polluted areas, also merit further research. The possibilities of levying compensation as part of an air pollution permit fee system require further definition and investigation, as do the advantages and disadvantages of a genuine government department for the environment. The same is true of several of the suggestions to improve the land use planning system such as the basis on which a third party right of appeal might operate. The advantages, disadvantages and feasibility of financing pressure groups to enable them to pursue environmental issues (including air pollution) in the community interest also demand investigation. Overcoming the difficulties encountered by such groups under present arrangements would be one more useful way of fulfilling the purpose of this study: planning pollution prevention.

NOTES

1 A document along the lines of that published on environmental impact assessment in the UK (Clark et al, 1981) is envisaged.
2 Many local authorities now permit oral presentations. The amount of information provided to the public within the planning system has improved markedly in recent years.
3 *The Town and Country Planning General Development Order 1988* SI 1988 No. 289.
4 *Town and Country Planning (Use Classes) Order 1987* SI 1987 No. 764.

Bibliography

Abt. Associates, Inc. (1977). *Integration of Environmental Considerations in the Comprehensive Planning and Management Process* Contract H-2175R, Office of Policy Development and Research, US Department of Housing and Urban Development, Washington, DC.

Ackerman, B. A., Hassler, W. T. (1981). *Clean Air/Dirty Coal.* Yale University Press, New Haven, CT.

Air Pollution Control Association (1988). *1988 Directory: Governmental Air Pollution Control Agencies.* APCA, Pittsburgh, PA.

American Law Institute. (1975). *A Model Land Development Code.* ALI, Washington, DC.

Arnold, G., Edgerley, E. (1967). Urban development in air pollution basins – an appeal to the planners for help. *Journal of the Air Pollution Control Association,* 17, 235–7.

Ashby, E., Anderson, M. (1981). *The Politics of Clean Air.* Oxford University Press.

Asimow, M. (1983). Delegated legislation: United States and United Kingdom. *Oxford Journal of Legal Studies,* 3, 253–276.

Babcock, R. F. (1966). *The Zoning Game.* University of Wisconsin Press, Madison, WI.

Babcock, R. F., Siemon, C. L. (1985). *The Zoning Game Revisited.* Oelgeshlager, Gunn and Hain, Boston, MA.

Bach, W. (1972). Urban climate, air pollution and planning, in Detwyler, T. R. and Marcus M. R. (eds) *Urbanization and Environment.* Duxbury Press, Belmont, CA.

Bachrach, P., Baratz, M. S. (1970). *Power and Poverty.* Oxford University Press, New York, NY.

Badaracco, J. L. (1985). *Loading the Dice: a Five-Country Study of Vinyl Chloride Regulation.* Harvard Business School Press, Boston, MA.

Bardach, E. (1977). *The Implementation Game: What Happens After a Bill Becomes Law.* MIT Press, Cambridge, MA.

Bardach, E., Kagan, R. A. (1982). *Going by the Book: The Problem of Regulatory Unreasonableness.* Temple University Press, Philadelphia, PA.

Barrett, M., Bisset, R., Chapman, K., Clark, B. D., Wathorn, P. (1981). *A Manual for the Assessment of Major Development Proposals.* Department of the Environment, HMSO.

Begg, D., Fischer, S., Dornbusch, R. (1984). *Economics.* McGraw-Hill.

Bingham, G. (1986). *Resolving Environmental Disputes: a Decade of Experience.* Conservation Foundation, Washington, DC.
Blowers, A. T. (1980). *The Limits of Power.* Pergamon Press.
Blowers, A. T. (1982). Much ado about nothing? – a case study of planning and power, in Healey, P., McDougall, G. and Thomas, M. J. (eds) *Planning Theory – Prospects for the 1980's.* Pergamon Press.
Blowers, A. T. (1984). *Something in the Air: Corporate Power and the Environment.* Harper and Row.
Bosselman, F. P., Callies, D. L. (1972). *The Quiet Revolution in Land Use Control.* Council on Environmental Quality, US Government Printing Office, Washington, DC.
Bosselman, F. P, Callies, D. L., Banta, J. S. (1973). *The Taking Issue,* Council on Environmental Quality, US Government Printing Office, Washington, DC.
Bosselman, F. P., Feurer, D. A., Siemon, C. L. (1977). *The Permit Explosion.* Urban Land Institute, Washington, DC.
Brady, G. L., Bower, B. T. (1982). Effectiveness of the US regulatory approach to air quality management : stationary sources. *Policy Studies Journal,* **11**, 66–76.
Brail, R. K. (1975). Land use planning strategies for air quality maintenance, in Roberts, J. J. (ed) *Proceedings of the Speciality Conference on Long Term Maintenance of Clean Air Standards, Chicago.* Air Pollution Control Association, Pittsburg, PA.
Branch, M. C., Leong, E. Y. (1972). *Air Pollution and City Planning.* Department of Environmental Science and Engineering, University of California, Los Angeles, CA.
Brickman, R., Jasanoff, S., Ilgen, T. (1985). *Controlling Chemicals : The Politics of Regulation in Europe and the United States.* Cornell University Press, Ithaca, NY.
Bugler, J. (1972). *Polluting Britain.* Penguin.
Burchell, R. W., Listokin, D. (eds). (1975). *Future Land Use: Energy, Environmental and Legal Constraints.* Center for Urban Policy Research, Rutgers University, New Brunswick, NJ.
Bureau of the Census (1986) *Statistical Abstract of the United States 1987.* US Government Printing Office, Washington, DC.
Byrne, A. (1986). *Local Government in Britain.* Penguin, 4th edition.
Caldwell, L. K., Hayes, L. R., MacWhirter, I. M. (1976) *Citizens and the Environment.* Indiana University Press, Bloomington, IN.
Callies, D. L. (1980). 'The Quiet Revolution' revisited. *American Planning Association Journal,* **46**, 135–144.
Callies, D. L. (1981). Public participation in the United States. *Town Planning Review,* **52**, 286–296.
Callies, D. L. (1984). *Regulating Paradise: Land Use Controls in Hawaii.* University of Hawaii Press, Honolulu, HI.
Callies, D. L. (1985). The taking issue revisited. *Land Use Law (July)* 6–8.
Canter, L. W. (1977). *Environmental Impact Assessment.* McGraw Hill, New York, NY.
Canter, L. W. (1983). A review of recent research on the utility of environmental impact assessment. *Proceedings, Symposium on Environmental Impact Assess-*

ment, Chania, April 1983. Centre for Environmental Management and Planning, University of Aberdeen, Aberdeen.
Central Directorate on Environmental Pollution. (1982). *Air Pollution Control*. Pollution Paper **18**, Department of the Environment, HMSO.
Central Statistical Office. (1988). *Annual Abstract of Statistics 1988*. HMSO, London.
Chandler, T. J. (1976). *Urban Climatology and its Relevance to Urban Design*. Technical Note 149, WMO 438, World Meteorological Organisation, Geneva.
Clawson, M., Hall, P. (1973). *Planning and Urban Growth: an Anglo-American Comparison*. Resources for the Future, Johns Hopkins University Press, Baltimore, MD.
Cohn, S., Elfers, K., Hufschmidt, M. M., Kaiser, E. J., Reichert, P. A., Stanland, R. E. (1974). *Promoting Environmental Quality through Urban Planning and Controls*. Report EPA – 600/5, 73–015, US Government Printing Office, Washington, DC.
Commission on Energy and the Environment. (1981). *Coal and the Environment*. Department of the Environment, HMSO.
Commission of the European Communities. (1977). European Community Policy and Action Programme on the Environment for 1977–1981. *Official Journal of the European Communities* **C139**. 13 June 1977.
Commission of the European Communities. (1979). *State of the Environment: Second Report*. CEC, Brussels.
Commission of the European Communities. (1985). Council directive of 27 June 1985 on the assessment of the effects of certain public and private projects on the environment. *Official Journal of the European Communities* **L175** 40–49, 5 July 1985.
Committee on Environment and Public Works, United States Senate (1981) *Hearings on Clean Air Act Oversight* 6 vols, 97th Congress, 1st Session, Report 97–1412, US Government Printing Office, Washington, DC.
Committee on Environment and Public Works, United States Senate. (1982). *Clean Air Act Amendments of 1982*. 97th Congress, 2nd Session, Report 97–666, US Government Printing Office, Washington, DC.
Committee on Environment and Public Works, United States Senate. (1987). *Clean Air Standards Attainment Act of 1987: Report of the Committee*. 100th Congress, 1st Session, Report 100–231, US Government Printing Office, Washington, DC.
Conservation Foundation (1982). *State of the Environment 1982*. CF, Washington, DC.
Conservation Foundation. (1984). *State of the Environment: an Assessment at Mid-Decade*. CF, Washington, DC.
Conservation Foundation. (1987). *State of the Environment: A View toward the Nineties*. CF, Washington, DC.
Corden, C. (1977). *Planned Cities: New Towns in Britain and America*. Sage.
Cotgrove, S. (1982). *Catastrophe or Cornucopia*. Wiley.
Council on Environmental Quality. (1970). *Environmental Quality 1970: 1st Annual Report*. US Government Printing Office, Washington, DC.
Council on Environmental Quality. (1974). *Environmental Quality 1974: 5th Annual Report*. US Government Printing Office, Washington, DC.

Council on Environmental Quality. (1978). Regulations for Implementing the Procedural Provisions of the National Environmental Policy Act, 40 *CFR* 1500–1508.
Council on Environmental Quality. (1981). *Environmental Trends*. CEQ, Washington, DC.
Council on Environmental Quality (1982a). *Environmental Quality 1981: 12th Annual Report*. US Government Printing Office, Washington, DC.
Council on Environmental Quality (1982b). 'Implementation of CEQ regulations on NEPA'. Document dated 12 July 1982, CEQ, Washington DC.
Council on Environmental Quality. (1984). *Environmental Quality 1983: 14th Annual Report*. US Government Printing Office, Washington, DC.
Council on Environmental Quality. (1987). *Environmental Quality 1985: 16th Annual Report*. US Government Printing Office, Washington, DC.
Craxford, S. R. (1976). Town and country planning, in Suess, M. J. and Craxford, S. R. (eds) *Manual on Urban Air Quality Management*. WHO Regional Publications, European Series 1, World Health Organisation, Copenhagen.
Craxford, S. R., Weatherley, M-L. P. M. (1966). Planning for clean air. *Journal of the Town Planning Institute*, 54, 158–171.
Crenson, M. A. (1971). *The Un-politics of Air Pollution*. Johns Hopkins Press, Baltimore, MD.
Culhane, P. J., Friesema, K. P., Beecher, D. (1987). *Forecasts and Environmental Decision-making: The Content and Predictive Accuracy of Environmental Impact Statements*. Westview Press, Boulder, CO.
Cullingworth, J. B. (1985). *Town and Country Planning in Britain*. Allen and Unwin, 9th edition.
Dear, M. J., Long, J. (1978). Community strategies in locational conflict, in Cox, K. (ed) *Urbanization and Conflict in Market Societies*. Methuen.
DeGrove, J. M. (1984). *Land, Growth, and Politics*. Planners Press, American Planning Association, Chicago, IL.
DeGrove, J. M., Stroud, N. E. (1987). State land planning and regulation: innovative roles in the 1980s and beyond. *Land Use Law [March]* 3–8.
Delafons, J. (1969). *Land-Use Controls in the United States*. MIT Press, Cambridge, MA, 2nd edition.
Department of the Environment. (1972). *Planning and Clean Air*. Draft Circular, DOE, London.
Department of the Environment. (1974). *110th Annual Report on Alkali, etc, Works 1973*. HMSO.
Department of the Environment. (1975). *111th Annual Report on Alkali, etc, Works 1974*. HMSO.
Department of the Environment. (1976). *Control of Smells from the Animal Waste Processing Industry*. Circular 43/76, HMSO.
Department of the Environment. (1980). *Development Control – Policy and Practice*. Circular 22/80, HMSO.
Department of the Environment. (1981). *Clean Air*. Circular 11/81, HMSO.
Department of the Environment. (1983). *Planning Gain – Obligations and Benefits which extend beyond the Development for which Planning Permission has been sought*. Circular 22/83, HMSO.

Department of the Environment. (1984). *Memorandum on Structure and Local Plans*. Circular 22/84 HMSO.
Department of the Environment. (1985a). *The Use of Conditions in Planning Permissions*. Circular 1/85, HMSO.
Department of the Environment. (1985b). *Development and Employment*. Circular 14/85, HMSO.
Department of the Environment. (1986). *Air Pollution Control in Great Britain: Review and Proposals*. Consultation Paper, DOE, London.
Department of the Environment. (1987a). *Simplified Planning Zones*. Circular 25/87, HMSO.
Department of the Environment. (1987b). *Digest of Environmental Pollution and Water Statistics*, 10. HMSO.
Department of the Environment. (1987c). *Development Control Statistics: England 1983/84–1985/86*. DOE, London.
Department of the Environment. (1988). *Environmental Assessment*. Circular 15/88, HMSO.
De Santo, R. S., Smith, W. H., Miller, J. A., McMillen, W. P., McGregor, K. A. (1976). *Open Space as an Air Resource Management Measure, 1*. EPA 450/3 – 76–028a, Environmental Protection Agency, Research Triangle Park, NC.
Domenici, P. (1979). Clean Air Act Amendments of 1977. *Natural Resources Journal*, 19, 475–485.
Douglas, M., Wildavsky, A. (1982). *Risk and Culture*. University of California Press, Berkeley, CA.
Downing, P. B. (1982). Cross-national comparisons in environmental protection: introduction to the issues. *Policy Studies Journal*, 11, 39–43.
Downing, P. B., Hanf, K. (1983a). *International Comparisons in Implementing Pollution Laws*. Kluwer Nijhoff, Dordrecht.
Downing, P. B., Hanf, K. (1983b). Modeling environmental regulation, in Downing and Hanf (1983a).
Drayton, W. (1981). Getting smarter about regulation. *Harvard Business Review*, 60(4), 38–52.
Ducsik, D. (1983). *Electricity Planning and the Environment*. Ballinger, Cambridge, MA.
Duerkson, C. J. (1982). *Dow vs California: a Turning Point in the Envirobusiness Struggle*. Conservation Foundation, Washington, DC.
Duerkson, C. J. (1983). *Environmental Regulation of Industrial Plant Siting*. Conservation Foundation, Washington, DC.
Enloe, C. H. (1975). *The Politics of Pollution in a Comparative Perspective*. David McKay, New York, NY.
Environmental Law Institute. (1981). *NEPA in Action: Environmental Offices in 19 Federal Agencies*. Council on Environmental Quality, Washington, DC.
Environmental Protection Agency. (1982). *Streamlining the Environmental Permitting Process: a Survey of State Reforms*. EPA, Washington, DC.
Environmental Protection Agency. (1988). *National Air Quality and Emissions Trends Report, 1986*. EPA, Research Triangle Park, NC.
Environmental Research and Technology, Inc. (1980). *The Impact of Air Quality Permit Procedures on Industrial Planning and Development*. Business Roundtable Air Quality Project, ERT, Concord, MA.

Environmental Research and Technology Inc. (1982). *Handbook on Requirements for Industrial Facilities under the Clean Air Act*. ERT, Concord, MA.
Fairfax, S. K., Ingram, H. M. (1981). The United States experience, in O'Riordan, T. and Sewell, W. R. D. (eds) *Project Appraisal and Policy Review*. Wiley.
Fisher, R., Ury, W. (1983). *Getting to Yes*. Penguin.
Fix, M., Muller, T. (1982). *The Impact of Regulation on Housing Costs*. Report 1342-1, Urban Institute, Washington, DC.
Foley, D. L. (1963). *Controlling London's Growth: Planning the Great Wen 1940–1960*. University of California Press, Berkeley, CA.
Frieden, B. J. (1979). *The Environmental Protection Hustle*. MIT Press, Cambridge, MA.
Friedlaender, A. F. (ed). (1978). *Approaches to Controlling Air Pollution*. MIT Press, Cambridge, MA.
Fudge, C., Barrett, S. (1981). Reconstructing the field of analysis, in Fudge, C., Barrett, S. (eds). *Policy and Action*. Methuen.
Garner, J. F., Callies, D. L. (1972). Planning law in England and Wales and in the United States. *Anglo-American Law Review*, 1, 292–334.
Gladwin, T. N. (1980). Patterns of environmental conflict over industrial facilities in the United States, 1970–78. *Natural Resources Journal*, 20, 243–274.
Godschalk, D. R., Brower, D. J., McBennett, L. D., Vestal, B. A. (1977). *Constitutional Issues of Growth Management*. ASPO Press, American Society of Planning Officials, Chicago, IL.
Greer, S. (1962). *Governing the Metropolis*. John Wiley, New York, NY.
Gregory, R. (1971). *The Price of Amenity*. Macmillan.
Griffith, J. A. G. (1966). *Central Departments and Local Authorities*. Allen and Unwin.
Haar, C. M. (1964). Comparisons and contrasts, in Haar, C. M. (ed). *Law and Land: Anglo-American Planning Practice*. Harvard University Press and MIT Press, Cambridge, MA.
Haar, C. M. (ed). (1984). *Cities, Law and Social Policy: Learning from the British*. Lexington Books, Lexington, MA.
Hagevik, G. H. (1970). *Decision-Making in Air Pollution Control*. Praeger, New York, NY.
Hagevik, G. H., Mandelker, D. R., Brail, R. K. (1974). *Air Quality Management and Land Use Planning*. Praeger, New York, NY.
Hagman, D. G. (1975). Commentary – land use controls: emerging and proposed reforms, in Burchell and Listokin (1975).
Haigh, N. (1987). *EEC Environmental Policy and Britain*. Longman, 2nd edition.
Hall, P. (1977). Urban and regional planning in Britain and America: ends and means. *Planning Outlook*, 20, 19–22.
Hall, P. (1982). *Great Planning Disasters*. University of California Press, Berkeley, CA.
Hall, P. (1983). The Anglo-American connection: rival rationalities in planning theory and practice, 1955–1980. *Environment and Planning* B 10 41–46.
Hall, P., Gracey, H., Prewett, R., Thomas, R. (1973). *The Containment of Urban England*. 2. Allen and Unwin.
Hart, S. L., Enk, G. A. (1980). *Green Goals and Greenbacks: State-Level Environmental Programs and their Associated Costs*. Westview Press, Boulder, CO.

Health and Safety Executive. (1978). *Industrial Air Pollution 1976*. HMSO.
Health and Safety Executive. (1982). *Industrial Air Pollution 1981*. HMSO.
Health and Safety Executive. (1984). *Industrial Air Pollution 1982*. HMSO.
Health and Safety Executive. (1987). *Industrial Air Pollution 1986*. HMSO.
Healy, R. G. (ed). (1978). *Protecting the Gold Shore*. Conservation Foundation, Washington, DC.
Healy, R. G. (1982). *America's Industrial Future: an Environmental Perspective*. Conservation Foundation, Washington, DC.
Healy, R. G., Rosenberg, J. S. (1979). *Land Use and the States*. Resources for the Future, Johns Hopkins University Press, Baltimore, MD.
Heidenheimer, A. J., Heclo, H., Adams, C. T. (1975) *Comparative Public Policy*. St Martin's Press, New York, NY.
Her Majesty's Government (1985). *Lifting the Burden*. Cmnd 9571, HMSO.
Her Majesty's Inspectorate of Pollution. (1988). *Best Practicable Means: General Principles and Practice*. BPM1, DOE, London.
Hill, A. C. (1971). Vegetation: a sink for atmospheric pollutants. *Journal of the Air Pollution Control Association*, 21, 341–346.
Hill, M (1983). The role of the British Alkali and Clean Air Inspectorate in air pollution control, in Downing and Hanf. (1983a).
Holdgate, M. W. (1979). *A Perspective of Environmental Pollution*. Cambridge University Press.
Holling, C. S. (ed). (1978). *Adaptive Environmental Assessment and Management*. Wiley.
House of Lords (1981). *Environmental Assessment of Projects*. Select Committee on the European Communities, 11th Report, Session 1980–81, HMSO.
Indiana University. (1983). *A Study of Ways to Improve the Scientific Content and Methodology of Environmental Impact Analysis*. Grant PRA-79-10014, National Science Foundation, Washington, DC.
International Union for Conservation of Nature and Natural Resources. (1980). *World Conservation Strategy*. IUCN, Gland, Switzerland, on behalf of the United Nations Environment Programme and the World Wildlife Fund.
Janelle, D. G. (1977). Structural dimensions in the geography of locational conflicts. *Canadian Geographer*, 21, 311–328.
Johnson, W. C. (1984). Citizen participation in local planning in the UK and USA: a comparative study. *Progress in Planning*, 21, 149–221.
Jones, C. O. (1975). *Clean Air: the Policies and Politics of Pollution Control*. University of Pittsburg Press, Pittsburg, PA.
Jowell, J. (1977). Bargaining in development control. *Journal of Planning and Environment Law [1977]*, 414–433.
Kelman, S. (1981). *Regulating America, Regulating Sweden: a Comparative Study of Occupational Safety and Health Policy*. MIT Press, Cambridge, MA.
Keogh, G. (1985). The economics of planning gain, in Barrett, S., Healey, P. (eds) *Land Policy: Problems and Alternatives*. Gower.
Keyes, D. L. (1976). *Land Development and the Natural Environment: Estimating Impacts*. Urban Institute, Washington, DC.
Kimber, R., Richardson, J. J. (eds). (1974). *Campaigning for the Environment*. Routledge and Kegan Paul.
Knödgen, G. (1983). Environmental regulations and the location of industry in

Europe. Paper to the Conservation Foundation Conference on Industrial Siting, San Francisco, Conservation Foundation, Washington, DC.

Knoepfel, P., Weidner, H. (1982). Formulation and implementation of air quality programmes: patterns of interest consideration. *Policy and Politics*, 10, 85–109.

Knoepfel, P., Weidner, H. (1983). Implementing air quality control programs in Europe: some results of a comparative study, in Downing and Hanf (1983a).

Kouwenhoven, J. A. (1961). *The Beer Can by the Highway*. Doubleday, Garden City, NY.

Kunreuther, H. C., Linnerooth, J. (1983). *Risk Analysis and Decision Processes: the Siting of Liquefied Energy Gas Facilities in Four Countries*. Springer-Verlag, Berlin.

Kurtzweg, J. A. (1973). Urban planning and air pollution control: a review of selected recent research. *Journal of the American Institute of Planners*, 39, 82–92.

Ledger, M. J. (1982). *An Assessment of the Effectiveness of Land Use Planning Powers to Control Pollution*. Unpublished PhD Thesis, University of Manchester.

Leonard, H. J. (1982). Environmental regulations, multinational corporations and industrial development in the 1980's. *Habitat International*, 6, 323–341.

Leonard, H. J. (1983). *Managing Oregon's Growth*. Conservation Foundation, Washington, DC.

Liroff, R. A. (1980). *Air Pollution Offsets: Trading, Selling and Banking*. Conservation Foundation, Washington, DC.

Liroff, R. A. (1981). NEPA legislation in the 1970s: a deluge or a dribble? *Natural Resources Journal*, 21, 315–330.

Liroff, R. A. (1986). *Reforming Air Pollution Regulation: The Toil and Trouble of EPA's Bubble*. Conservation Foundation, Washington, DC.

Lowe, P., Goyder, J. (1982). *Environmental Groups in Politics*. Allen and Unwin.

Lundquist, L. J. (1978). The comparative study of environmental politics: from garbage to gold? *International Journal of Environmental Studies*, 12, 89–97.

Lundquist, L. J. (1980). *The Hare and the Tortoise: Clean Air Policies in the United States and Sweden*. University of Michigan Press, Ann Arbor, MI.

Lyday, N. (1976). *The Law of the Land*. Urban Institute, Washington, DC.

Magazine, A. H. (1977). *Environmental Management in Local Government*. Praeger, New York, NY.

Magorian, C., Morell, D., (1982). *Siting Hazardous Waste Facilities*. Ballinger, Cambridge.

Majone, G., Wildavsky, A. (1979). Implementation as evaluation, in Pressman, J. L., Wildavsky, A. (eds). *Implementation*. University of California Press, Berkeley, CA, 2nd edition.

Manners, I. R., Rudzitis, G. (1982). Federal air quality legislation: implications for land use, in Hoffman, G. (ed). *Federalism and Regional Development*. University of Texas Press, Austin, TX.

Masser, I. (1986). Some methodological considerations, in Masser and Williams (1986).

Masser, I., Williams, R. (1986). *Learning from Other Countries: the Cross-National Dimension in Urban Policy-Making*. Geo Books.

Massey, D., Catalano, A. (1978). *Capital and Land: Land Ownership by Capital in Great Britain*. Edward Arnold.

Mazmanian, D. A., Sabatier, P. A. (1981). The implementation of public policy: a framework for analysis, in Mazmanian, D. A., Sabatier, P. A. *Effective Policy Implementation.* Lexington Books, Lexington, MA.

McAllister, D. M. (1980). *Evaluation in Environmental Planning.* MIT Press, Cambridge, MA.

McCormick, R. A. (1971). Air pollution in the locality of buildings. *Philosophical Transactions of the Royal Society, London A.* **269**, 515–526.

McLoughlin, J., Foster, M. J. (1982) *Law and Practice Relating to Pollution Control in the United Kingdom.* Graham & Trotman.

Melosi, M. V. (ed). (1980). *Pollution and Reform in American Cities 1870–1930.* University of Texas Press, Austin, TX.

Meyerson, M., Banfield, E. C. (1955). *Politics, Planning and the Public Interest.* Free Press, New York, NY.

Miller, C., Wood, C., McLoughlin, J. (1980). *Land Use Planning and Pollution Control IV.* Pollution Research Unit, University of Manchester.

Miller, C., Wood, C. (1983). *Planning and Pollution.* Oxford University Press.

Ministry of Housing and Local Government (1967). *103rd Annual Report on Alkali, etc, Works 1966.* HMSO.

Mitchell, B. (1979). *Geography and Resource Analysis.* Longman.

Mogulof, M. B. (1971). *Governing Metropolitan Areas.* Urban Institute, Washington, DC.

Morell, D. (1974). Air quality and land use, in Hussey, E. T. (ed). *Proceedings of the Conference on Air Quality Impact Analysis for Application in Land Use and Transportation Planning, Berkeley.* University Extension, University of California, Berkeley, CA.

Morell, D., Singer, G. (eds). (1980). *Refining the Waterfront.* Oelgeschlager, Gunn and Hain, Cambridge, MA.

Moss, E. (ed). (1977). *Land Use Controls in the United States.* Natural Resources Defense Council, Dial Press/James Wade, New York, NY.

Munn, R. E. (1983). The atmospheric component of environmental impact assessment, in PADC Environmental Impact Assessment and Planning Unit (ed) (1983). *Environmental Impact Assessment.* Martinus Nijhoff, Dordrecht.

National Academy of Sciences. (1981). *On Prevention of Significant Deterioration of Air Quality.* National Academy Press, Washington, DC.

National Commission on Air Quality. (1981). *To Breathe Clean Air.* NCAQ, Washington, DC.

Noble, J. H., Banta, J. S., Rosenberg, J. S. (1977). *Groping through the Maze.* Conservation Foundation, Washington, DC.

Office of Population Census and Surveys. (1983). *Census 1981: Country of Birth: Great Britain.* HMSO, London.

O'Hare, M., Bacow, L., Sanderson, D. (1983). *Facility Siting and Public Opposition.* Van Reinhold Nostrand, New York, NY.

Organisation for Economic Co-operation and Development. (1975). *The Polluter-Pays Principle.* OECD, Paris.

Organisation for Economic Co-operation and Development. (1987). *OECD Environmental Data Compendium 1987.* OECD, Paris.

O'Riordan, T., D'Arge, R. C. (eds). (1979). *Progress in Resource Management and Environmental Planning. 1.* Wiley.
O'Riordan, T. (1979a). The role of environmental quality objectives in politics of pollution control, in D'Arge and O'Riordan (1979).
O'Riordan, T. (1979b). Public interest environmental groups in the United States and Britain. *American Studies*, **13**, 409–438.
Peacock, A. (ed.) (1984). *The Regulation Game: How British and West German Companies Bargain with Government.* Blackwell.
Pearce, D., Edwards, L., Beuret, G. (1979). *Decision Making for Energy Futures: a Case Study of the Windscale Inquiry.* Macmillan.
Pearlman, K. (1977). State environmental policy acts: local decision making and land use planning. *Journal of the American Institute of Planners*, **43**, 42–53.
Pelham, T. G. (1979). *State Land-Use Planning and Regulation.* Lexington Books, Lexington, MA.
Petts, J., Hills, P. (1982). *Environmental Assessment in the UK.* Institute of Planning Studies, University of Nottingham.
Photochemical Oxidants Review Group. (1987). *Ozone in the United Kingdom.* Department of the Environment, London.
Popper, F. J. (1981). *The Politics of Land-Use Reform.* University of Wisconsin Press, Madison, WI.
Quarles, J. (1979). Federal regulation of new industrial plants. *Environmental Reporter*, **10**(1), 1–51. Monograph 28, Bureau of National Affairs, Washington, DC.
Raffle, B. J. (1978). The new Clean Air Act – getting clean and staying clean. *Environmental Reporter*, **8**(47), 1–28. Monograph 26, Bureau of National Affairs, Washington, DC.
Rees, J. A. (1985). *Natural Resources: Allocation, Economics and Policy.* Methuen.
Reilly, W. (1974). New directions in federal land use legislation, in Listokin, D. (ed). *Land Use Controls: Present Problems and Future Reform.* Center for Urban Policy Research, Rutgers University, New Brunswick, NJ.
Rhodes, G. (1981). *Inspectorates in British Government.* Allen and Unwin.
Ridker, R. G. (1967). *The Economic Costs of Air Pollution.* Praeger, New York, NY.
Roberts, J. J., S., Croke, E. J., Booras, S. (1975). A critical review of the effect of air pollution control regulations on land use planning. *Journal of the Air Pollution Control Association*, **25**, 500–520.
Rose, R. (ed). (1974). *The Management of Urban Change in Britain and Germany.* Sage.
Rosenbaum, N. (1976). *Land Use and the Legislatures.* Urban Institute, Washington, DC.
Royal Commission on Environmental Pollution. (1976). *Fifth Report: Air Pollution Control: an Integrated Approach.* Cmnd. 6371, HMSO.
Royal Commission on Environmental Pollution. (1985). *Eleventh Report: Managing Waste: the Duty of Care.* Cmnd 9675, HMSO.
Royal Commission on Environmental Pollution. (1988). *Twelfth Report: Best Practicable Environmental Option.* Cm 310, HMSO.
Royston, M. G. (1979). *Pollution Prevention Pays.* Pergamon Press.
Rydell, C. P., Schwarz, G. (1968). Air pollution and urban form: a review of current literature. *Journal of the American Institute of Planners*, **34**, 115–120.

Sabatier, P. A., Mazmanian, D. A. (1983). *Implementation and Public Policy*. Scott Freeman, Glenview, IL.
Saunders, P. J. W. (1976). *The Estimation of Pollution Damage*. Manchester University Press.
Saunders, P. J. W., Wood, C. (1974). Plants and air pollution. *Landscape Design*, 105, 28–30.
Scarrow, H. (1971). Policy pressures by British local government: the case of regulation in the 'public interest'. *Comparative Politics*, 4, 1–28.
Scorer, R. S. (1972). *Air Pollution*. Pergamon Press.
Seley, J. A. (1983). *The Politics of Public-Facility Planning*. Lexington Books, Lexington, MA.
Smith, P. J. (ed). (1975). *The Politics of Physical Resources*. Penguin.
Social Audit. (1974). *The Alkali Inspectorate*. SA, London.
Stafford, H. A. (1985). Environmental protection and industrial location. *Annals of the Association of American Geographers*, 75, 227–240
Stern, A. C. (ed). (1977a). *Air Pollutants, their Transformation and Transport, Air Pollution I*. Academic Press, New York, NY, 3rd edition.
Stern, A. C. (ed). (1977b). *The Effects of Air Pollution, Air Pollution II*. Academic Press, New York, NY, 3rd edition.
Stern, A. C. (ed). (1977c). *Engineering Control of Air Pollution, Air Pollution IV*. Academic Press, New York, NY, 3rd edition.
Sternlieb, G., Burchell, R. W., Hughes, J. W. (1975). An introduction to the invited papers, in Burchell and Listokin (1975).
Storper, M., Walker, R., Widess, E. (1981). Performance regulation and industrial location: a case study. *Environment and Planning*, 13, 321–338.
Strauss, A. (1978). *Negotiations*. Jossey-Bass, San Francisco, CA.
Symonds, W. (1982). Washington in the grip of the green giant. *Fortune*, 106 (7), October 4, 136–140, 151, October 4.
Talbot, A. R. (1983). *Settling Things: Six Case Studies in Environmental Mediation*. Conservation Foundation, Washington, DC.
Train, R. E. (1976). The EPA programs and land use planning. *Columbia Journal of Environmental Law*. 2, 255–289.
Van Meter, D., Van Horn, C. (1975). The policy implementation process, a conceptual framework. *Administration and Society*, 6, 445–488.
Venezia, R. A. (1976). Air quality management through land-use measures. *Journal of the Urban Planning and Development Division, Proceedings of the American Society of Civil Engineers*, 102, 95–103.
Vogel, D. (1983a). Comparative regulation: environmental protection in Great Britain. *The Public Interest*, 72, 88–106.
Vogel, D. (1983b). Comparing policy styles: environmental protection in the United States and Great Britain. *Public Administration Bulletin*, 42, 65–78.
Vogel, D. (1986). *National Styles of Regulation: Environmental Policy in Great Britain and the United States*. Cornell University Press, Ithaca, NY.
Voorhees, A. M. and Associates, Inc (1974). *Land Use and Transportation Considerations*. Guidelines for Air Quality Maintenance Planning and Analysis, 4, EPA – 450/4 – 74 – 004, Environmental Protection Agency, Research Triangle Park, NC.
Voorhees, A. M., and Associates, Inc and Ryckman, Edgerley, Tomlinson and Asso-

ciates. (1971). *A Guide for Reducing Air Pollution through Urban Planning.* NTIS PB 207 510, National Technical Information Service, Springfield, VA.
Webster, C. A. R. (1981). *Environmental Health Law.* Sweet and Maxwell.
West, R., Foot, P. (1975). Anglesey: aluminium and oil, in Smith (1975).
Weidner, H. (1987). *Clean Air Policy in Great Britain: Problem-Shifting as Best Practicable Means.* Wissenschafts Zentrum, Berlin.
Williams, R. A. (1960). *American Society: a Sociological Interpretation.* Knopf, New York, NY, 2nd edition.
Willis, K. (1982). Planning agreements and planning gain. *Planning Outlook*, **24**, 55–62.
Wilson, G. K. (1985). *The Politics of Safety and Health: Occupational Safety and Health in the United States and Britain.* Oxford University Press.
Wood, C. (1976). *Town Planning and Pollution Control.* Manchester University Press.
Wood, C. (1979). Land use planning and pollution control, in D'Arge and O'Riordan. (1979).
Wood C., Pendleton, N. (1979). *Land Use Planning and Pollution Control in Practice.* Occasional Paper 4, Department of Town and Country Planning, University of Manchester, Manchester.
Wood, C. (1982). The impact of the European Commission directive on environmental planning in the United Kingdom. *Planning Outlook*, **24**, 92–98.
Wood, C. M. (1988). EIA and BPEO: Acronyms for good environmental planning? *Journal of Planning and Environment Law [1988]*, 310–321.
Wood, C., Hooper, P. (forthcoming). The effects of the relaxation of planning controls in enterprise zones on industrial pollution. *Planning and Environment A.*
Wood C., Jenkins, T. (1985). Threat of pollution after abolition. *Town and Country Planning*, **54**, 366–367.
Wood, C. M. (Forthcoming). The North Carolina oil refinery. *Carolina Planning.*
World Bank. (1978). *Environmental Considerations for the Industrial Development Sector.* WB, Washington, DC.
World Health Organisation. (1987). *Air Quality Guidelines for Europe.* WHO Regional Publication, European Series 23, WHO Regional Office for Europe, Copenhagen.

Index

Absorption, 25, 99
Acid rain, 9, 19, 20, 34, 50, 52, 93, 97
Activated carbon filter, 25, 82, 182, 195, 210
Air pollution:
 control agency, 3, 14, 25, 148–52, 163, 168, 170, 179–83, 199, 211, 217, 219
  characteristics, 149–50, 180–1
 effects of, 16–21
 trends, 40–1, 95–7
Air Quality Act 1967 (US), 42
Air quality, 37, 41, 45, 121, 165
 control regions 42, 52, 151
Alabama, 70, 74
Alaska, 32
Alkali Inspectorate, 98; see also HMIP
Alkali, etc, Works Regulation Act 1906 (UK), 97, 99, 103, 104, 106, 121, 129
Amenity, 10, 13, 111, 127, 139, 157, 202, 203
 societies, 137, 164
American Planning Association, 61
Anglesey aluminium smelter, 139, 172
Animal treatment works, 97, 104
Appeal, 5, 12, 36, 82, 110, 112, 120, 127, 129, 150, 179, 182, 193, 195, 208, 211, 212, 218, 220, 221, 222
Area emission standards, 21, 33
Areas of outstanding natural beauty, 111, 113, 114
Asbestos, 20, 115, 145
Asphalt plant, 74–6, 145, 146, 153, 154, 155, 159, 161, 162, 171, 175, 176, 178, 180, 186, 187, 192, 194, 198, 200, 206, 207

Atrazine, 129, 130
Attitude, 5, 143, 154, 159, 160, 214; see also Developer attitude
Authorisation process, 166–7, 196–8, 212, 214, 218, 222

Ballot, 59, 84, 85, 167, 193, 195
Bargaining, 147, 149, 151, 152, 157, 162, 166, 177, 179, 196, 210, 218; see also Negotiation
Beaver Committee, 98
Benefits of regulation/control, 13, 57, 203, 207, 213, 219
Best available control technology (BACT), 45, 46, 47, 181, 217
Best practicable environmental option, 9, 98, 222
Best practicable means (BPM), 9, 45, 99, 101, 102, 105, 107, 120, 121, 132, 150, 177, 181, 222
Bedfordshire brickworks, 109, 155
Bolton, 126–8, 130–2
Boston, 52
Bronchitis, 20, 98
Brunswick County (North Carolina), 76–8
Brunswick Energy Company (BECO), 76–8
Bubble provisions, 44, 49, 51, 86, 87, 182
Buffer zone, 27, 28, 30, 55, 76, 78, 131, 132, 159, 190, 214
Building design and arrangement, 30, 214

California, 38, 54, 59, 66, 71, 85–9, 185, 217

Index   237

California Coastal:
    Act 1976, 59
    Commission, 53
California Environmental Quality Act
    1970, 85
Canada, 34
Cancer, 81, 83
Carbon:
    dioxide, 19
    monoxide, 18, 20, 41, 42, 43, 46, 49,
        84
Carcinogens, 81, 176
Carolina Coastal Crossroads, 76–8,
    161
Cecil County (Maryland), 81–3
Chemical:
    formulation plant, 128–30, 153, 175,
        178, 179, 180, 183, 199, 200, 201,
        206, 221
    process, 99, 115, 128
    production facility, 87–9, 145, 146,
        149, 158, 161, 167, 176, 180, 181,
        190, 192, 193, 197, 198, 208
Cheshire, 133–5
Chester, 133
Chevron Inc., 85–7, 179, 182
Chicago, 167
Childs (Maryland), 81–3
Chimney height, 9, 26, 28, 29, 99, 102,
    103, 107, 120, 121, 123, 124, 128,
    135, 136, 139, 151, 159, 168, 182,
    200, 211
Chlorides, 20, 25
Chloride Ltd, 130–2
Citizens for a Better Environment,
    85–7, 161, 162, 165, 194, 195, 208
Citizens for Common Sense, 83, 146
City Planning Enabling Act 1966 (US),
    54
Clean Air Acts (UK), 98, 103, 104, 106,
    107, 121, 124
Clean Air Acts 1970 and 1977 (US), 37,
    41, 42, 45, 46, 47, 49–52, 61, 68,
    69, 70, 79, 89, 151, 209
Climate, 34, 93
Cluster zoning, 56, 185
Coastal management permit, 71, 74,
    187

Coastal Zone Management Act 1972
    (US), 58, 61
Colorado, 34
Combustion, 25, 103
Commission on Energy and the
    Environment, 107, 111
Commission of the European
    Communities (CEC), 1, 2, 24, 31, 98,
    107, 113, 114, 115, 118, 121, 138,
    196
Committee of Environment and Public
    Works of Senate, 51, 52
Community charge, 219
Compensation, 12, 53, 108, 110, 148,
    207, 213, 215, 216, 218, 219, 220,
    222
Complaints, 125, 132, 134, 136, 137,
    138, 169, 213
Compliance (with regulations), 39, 47,
    168, 169, 198, 209, 218
Comprehensive planning, 55, 59
Conditions, 3, 7, 27, 64, 80, 83, 105,
    110, 111, 114, 115, 121, 124, 125,
    126, 129, 130, 131, 134, 148, 151,
    158, 171, 183, 191, 200, 201, 205,
    206, 207, 211, 212, 218, 221
    nature of, 168–9, 198
Conflict, 140, 141, 143, 146, 156, 160,
    165, 166, 209, 212
Conservation area, 111, 113, 114
Conservation Foundation, 70
Construction permit, 44, 79
Consultation, 12, 27, 100, 115, 119,
    120, 123, 125, 136, 152, 159, 165,
    166, 183, 190, 195, 211, 214, 221
Contra Costa County (California),
    87–9
Control of Pollution Act 1974 (UK), 98
Corporatism, 197, 203
Corps of Engineers, 77, 78, 80
Corrosion, 20
Cost-effectiveness, 12, 38, 203, 204,
    220
Costs:
    of control, 1, 12, 42, 50, 106, 147,
        152, 157, 159, 161, 179, 182, 187,
        188, 203, 206
    of damage, 1, 20, 153, 206, 213

Costs – *continued*
  to the public, 153, 204, 213
Council on Environmental Quality
  (CEQ), 8, 24, 63, 68
Council for the Protection of Rural
  England, 201
Courts, 7, 42, 53, 56, 76, 78, 82, 87,
  95, 101, 150, 161, 163, 178, 182,
  192, 193, 195, 197, 199, 204, 211
Creosote storage facility, 71–4, 144,
  145, 149, 151, 154, 161, 163, 164,
  167, 175, 176, 178, 181, 186, 187,
  192, 193, 198, 206
Cremer and Warner, 133
Crewe Chemicals Ltd, 128–30
Criteria pollutants, 43

Dade county (Florida), 78–80
Damage from pollution, 1, 12, 20, 21,
  27, 102, 104, 133, 137, 145, 151,
  156, 164, 168, 176, 182, 188, 194,
  205, 207, 219
Department of the Environment (DOE),
  8, 94, 98, 100, 107, 123, 178
  for Northern Ireland, 94
Department of Health and Social
  Security, 131
Derelict land, 109, 126
Desulphurisation, 9
Developer, 3, 5, 7, 14, 143–8, 173–9,
  198, 199, 205, 211, 216, 219
  acceptability to, 169–70, 198–9
  attitude, 143, 144, 175, 198, 200
  characteristics, 144–5, 173–5, 210
  influence, 144, 175
  reputation, 144, 175
Development:
  control, 2, 108, 109–11, 119, 164
  plans, 109, 113, 178, 186
Dioxin, 83, 84, 145, 154, 176
Discontinuance, 111, 136, 137, 171
Discretionary powers, 92, 101, 167,
  186, 211
Dispersion of pollution, 22, 26, 29
District heating, 28, 31
Donora (Pennsylvania), 41
Dose/response function, 20, 23
Dow Chemical Co., 87–9
Dumping at sea, 127

Dust, *see* Particulate matter
Dye works, 97

Education, 7, 11, 27, 28, 29, 35, 216
Effectiveness, 8, 11, 12, 13, 159, 198,
  200, 203, 204, 209–10, 216
Efficiency, 11, 12, 13, 38, 92, 204–6
Electrostatic precipitation, 25, 84, 99,
  137, 168
Ellesmere Port and Neston, 133, 134
Emission:
  banking, 51, 196
  density zoning, 23, 70
  standards, 22–3, 100, 101, 148, 149,
  158, 169
  trading, 49, 51, 52
Employment, 10, 83, 87, 94, 113, 153,
  159, 207
Emphysema, 20, 98
Energy recovery facility, 83–5, 145,
  146, 147, 152, 154, 155, 158, 159,
  161, 163, 164, 165, 167, 168, 176,
  177, 188, 191, 192, 193, 195, 198,
  201, 205, 206, 208, 210, 217
Enforcement, 7, 13, 39, 42, 49, 52, 59,
  60, 80, 105, 107, 111, 129, 130,
  135, 136, 138, 148, 160, 168,
  170–1, 197, 199–201, 213, 218, 222
Enterprise zones, 108, 112, 123
Environmental:
  groups, 37, 38, 39, 65, 161, 163,
  189, 194, 201, 202, 212, 217, 222
  management, 70, 113, 219
  regulation, 10, 37–9, 95, 188, 202,
  203
  standards, 8, 9, 12, 21–3, 37, 42
Environmental health
  departments/officers, 104, 106, 119,
  121, 123, 124, 125, 126, 127, 130,
  131, 132, 135, 138, 149, 150, 152,
  156, 159, 182, 206, 211
Environmental impact:
  assessment (EIA), 8, 9, 24–5, 31, 45,
  58, 63–6, 68, 113, 114–16, 127,
  129, 158, 162, 190, 212, 218, 221,
  223
  report, 77, 79, 85, 88, 116, 158, 204
  statement (EIS), 24, 64, 65, 66, 77,
  78, 116

Environmental Protection Agency (EPA), 35, 42, 50, 51, 64, 69, 76, 83, 84, 86, 89, 217, 220
Epidemiological studies, 77, 147
Equity, 11, 12, 204, 206–9
  outcome, 12, 206–7
  procedural, 12, 206, 207–8, 221
Euclidean zoning, 55
Explosions, 126, 131
External scrutiny, 150, 154, 177, 180, 186

Ferro-Alloys and Metals Ltd, 137–9
Fertiliser plant, 133–5, 145, 146, 154, 155, 156, 158, 162, 163, 164, 170, 179, 188, 190, 208
Filtration, 25, 99, 131, 168
Finding of no significant impact (FONSI), 64
Fines, 50, 198, 199, 213
First English Evangelical Lutheran Church, 53
Florida, 59, 71, 78–80
Forestry, 20, 92
Friends of the Earth, 88
Frodsham, 133
Frontier ethic, 33, 184
Fugitive emissions, 41, 100
Fumes, 20, 34, 103, 111, 125

Galaxy Chemicals Company, 81
Gary (Indiana), 167
Geddes, 108
General Development Order (UK), 111, 221
Glass fibre works, 123–5, 144, 159, 161, 168, 170, 180, 182, 183, 188, 190, 200
Glossop, 137–9
Government system, 35–7, 93–4
Greater Manchester, 125, 127, 131
Green belts, 111, 113, 114
Greenfield projects, 39, 140
Ground level concentration, 26, 99
Guidelines, 22, 100

Hawaii, 32, 54, 59
Hayes Group, 135–7

Health and Safety Executive (HSE), 98
Health, 18, 21, 37, 42, 103, 108, 137, 203
Helsby, 133
Her Majesty's Inspectorate of Pollution (HMIP), 50, 94, 98, 99–107, 114, 116, 119, 121, 123, 128, 130, 131, 132, 134, 135, 136, 137, 138, 139, 144, 149, 150, 156, 162, 170, 177, 180, 181, 182, 183, 191, 192, 211, 219, 220, 221
Herbicides, 128, 130
High Court, 110, 138, 161
High Peak Borough Council, 137–9
Holme Pierrepoint, 139
Hospitals, 28, 29, 120, 135, 136, 145
'Housekeeping' controls, 25, 100
Howard, 108
Hydrocarbons, 18, 20, 79, 88, 177

Illinois, 167
Implementation, 13, 14, 141, 143, 160, 167–71, 198–201, 209, 212, 218, 222
Improvement notice, 105, 138
Incinerator, 44, 46; *see also* Sewage sludge incinerator
Individualism, 34, 184,
Industrial:
  developments, 114, 120, 127, 129, 164
  processes, 44, 98, 99, 100, 115
Industry, 16, 28, 29, 34, 39, 54, 55, 78, 93, 113, 166, 202, 203
Information, 36, 162, 165, 166, 204, 207, 212, 215, 219, 220, 221, 222, 223
Institute of Occupational Hygiene, 132
Institution of Environmental Health Officers, 100
Interest groups; *see* Environmental groups
International Commission on Radiological Protection, 219
Inversion, 28, 34, 93

Kirklees, 128–30

Land use:
  controls, 2, 26–31, 34, 52–63, 69, 70, 82, 94, 108–14, 158, 208
  control techniques, 28–31, 160, 212, 214
  planning agency, 3, 5, 54, 57, 104, 126, 143, 151, 153–60, 163, 168, 170, 183–91, 199, 212, 217, 221
  characteristics, 153–6, 184–7
Law suits/litigation, 37, 38, 61, 65, 195, 198
Lead, 20, 43, 97, 98, 131, 160, 169
Lead battery plant, 130–2, 144, 147, 150, 154, 156, 158, 159, 168–9, 170, 175, 176, 177, 179, 180, 181, 190, 191, 198, 200, 205
Leathers Chemical Co. Ltd, 135–7
Legal framework, 41–3, 52–4, 97–104, 108–11
Liaison committee, 77, 106, 132
Little Elk Creek Civic Association, 81–3, 163, 165
Local government, 35, 36, 55, 56, 108, 176, 211, 216, 218
Local Government (Access to Information) Act 1985 (UK), 111
Local Government Act (1972) (UK), 127
Local Government Ombudsman, 126, 137, 138
Local revenues/tax, *see* Tax
Locally undesirable land uses, 165
London, 93
Los Angeles, 34, 52
Louisiana, 61, 70, 71–4
Lowest achievable emission rate (LAER), 47, 49, 50, 84, 181, 217

Maine, 59
Maryland, 59, 71, 81–3
Media attention, 37, 76, 85, 127, 163, 175, 183, 194, 195
Mediation, 153, 154, 213, 215, 216
Meteorological control, 25, 26
Metropolitan Dade County (Florida), 78–80
Metropolitan districts, 108, 219
Molybdenum smelter, 137–9, 144, 145, 147, 150, 157, 158, 159, 161, 168, 176, 182, 183, 188, 199, 200, 206

Metropolitan statistical areas (MSAs), 32, 36
Miami, 36, 78–80
Midland City/County (Texas), 74–6
Minnesota, 70
Mississippi, 74
Mitigation, 5, 24, 64, 65, 80, 85, 118, 127, 146–8, 151–2, 157–60, 161, 165–6, 175, 177–9, 182–3, 188–91, 194–5, 205, 207, 210, 215, 216, 218
Models, 22, 46, 65, 77, 85, 86, 217
Monitoring, 21, 23, 46, 60, 64, 77, 107, 129, 132, 133, 134, 135, 137, 138, 158, 169, 177, 178, 181, 211, 213, 217
Motor vehicles, 16, 42, 69, 70

National Commission on Air Quality, 21, 41, 50, 192
National Environmental Policy Act 1969 (NEPA) (US), 2, 24, 61, 63, 65
National parks, 33, 45, 92, 111, 113, 114
National Wildlife Federation, 161
Nature of the site, 145, 151, 156, 164, 176, 181, 187, 194
Negotiation, 3, 12, 86, 95, 141, 143, 152, 157, 159, 161, 178, 211, 213, 215, 216; *see also* Bargaining
New Hanover County (North Carolina), 76–8
New Jersey, 59
New source performance standards (NSPS), 22, 43, 44, 49, 90, 217
New Towns Act 1965 (UK), 123
New towns, 10, 58, 113, 123, 144, 158
New York, 36, 52, 63
Nitrogen oxides, 18, 20, 41, 43, 46, 49, 88, 98
Noise, 111, 114, 118, 123, 127, 130
Nonattainment, 47–9, 51, 52, 69, 79, 88, 181, 211
Non-combustion process, 103, 104
Non-conforming uses, 55, 75, 85
Non-registered stationary sources, 103–4, 118
North Carolina, 70, 76–8
North West Water Authority, 126–8

Index 241

Northern Ireland, 93, 94
Not in my back yard (NIMBY), 62, 165
Nuisance, 1, 52, 56, 103, 104, 118

Objectors, 3, 5, 7, 14, 130, 134, 138, 139, 143, 144, 156, 160–6, 172, 181, 191–5, 208, 211, 212, 218, 222
  access to decision-making, 162, 191
  characteristics, 160–3, 191–4
Odours, 56, 80, 82, 111, 114, 118, 124, 130, 133, 137, 168
Offsets, 43, 47, 49, 50, 69, 79, 84, 85, 86, 89, 149, 177, 211, 217, 220
Oil refinery, 27, 44, 46, 76–8, 85–7, 107, 144, 145, 146, 147, 150, 154, 157, 158, 159, 161, 162, 163, 164, 166, 175, 176, 179, 180, 181, 182, 190, 191, 192, 193, 194, 195, 196, 198, 199, 204, 205, 206, 208
'One stop' permit, 79, 196
Oregon, 34, 54, 59, 71, 83–5
Oregon City (Oregon), 83–5
Oregonians for Clean Air, 83–5
Ozone, 20, 41, 42, 43, 79, 84, 88, 203, 220

Palmer Barge Inc., 71–4
Participation (public), 36, 37, 44, 51, 60, 61, 109, 157, 162, 166, 191, 192, 205, 207, 209, 211, 213, 215
Particulate matter, 16, 19, 20, 23, 25, 41, 42, 43, 45, 46, 47, 49, 56, 79, 83, 84, 88, 103, 111, 121, 124, 125, 133, 134, 181
Peak District National Park, 137
Pennine Chemical Services Ltd, 130–2
Pennsylvania, 41
People for Open Space, 88
Personnel resources, 38, 70, 152, 158, 165, 183, 189, 199
Petaluma (California), 62
Petitions, 83, 138
Petroplex Land and Development Co., 74–6
Phenol, 124–6
Photochemical smog, 18, 20, 34, 93, 97
Physical characteristics, 32–4, 91–2
Pittsburg (California), 88, 89

Planned unit development, 55, 57, 185, 196
Planning:
  agreement, 111, 112, 114, 131–2, 133, 158
  conditions, *see* Conditions
  control techniques, *see* Land use control techniques
  gain, 57, 157, 189, 215
  permission, 95, 131, 138
Plans, 27, 52, 54, 78, 92, 108, 109, 113, 118, 119, 126, 154, 185, 186
'Plat', 57
Pluralism, 34, 89, 97
Policy context, 149, 153, 180, 185
Polluter pays principle, 12, 220
Pollution:
  control, 21, 27, 63, 94, 99, 109, 113, 137, 170, 190, 198
  definition, 14
  effects, 16–21
  process, 15, 16, 25, 26
  sources, 2, 18, 44
  type of, 145, 175
Population density, 32, 55, 91, 207
Portland, 83–5
Portland Metropolitan Service District (Metro), 83–5
Power plants/stations, 26, 37, 79, 80, 84, 112, 115, 158, 188
Power, 140, 144, 166
Powers, 7, 14, 149, 153, 180, 184
Practice and reform, 49–52, 61–3, 104–7, 112–14
Preconstruction review, 44, 45, 69, 90
Pressure groups, *see* Environmental groups
Presumptive limits, 22, 100, 116, 131
Prevention of significant deterioration (PSD), 43, 44–7, 49, 50, 51, 52, 68, 69, 77, 79, 83, 89, 182, 217
Priority to pollution control, 169, 199
Project:
  benefits, 146, 148, 176
  characteristics, 145–6, 151, 156–7, 164–5, 175–7, 181–2, 187–8, 194
  modification, 87, 137, 177, 187, 188
Property tax, *see* Tax

Providence (Maryland), 81–3
Public Health Act 1936 (UK), 97, 103, 104
Public hearing/inquiry, 12, 36, 54, 59, 75, 78, 84, 87, 88, 100, 109, 110, 112, 115, 127, 128, 129, 133, 134, 135, 136, 139, 172, 208, 212, 217, 218
Publishers Paper Co., 83–5

Quiet reforms/revolution, 57, 60, 215, 218

Ramapo (New York), 63
Reasonably available control technology (RACT), 44
Receptor, 5, 26, 28, 159, 211
Recreational facilities, 28, 29, 41, 54
Registered stationary services, 99–103
Residential:
  area, 54, 55, 97, 156, 159
  property, 7, 27, 28, 93, 102, 120, 123, 145, 148
Residents for Unpolluted Neighbourhoods, 81–3, 164, 165
Resin manufacturing mill, 125–6, 144, 152, 155, 171, 176, 187, 193
Resource Conservation and Recovery Act 1981 (US), 58
Resources, 70, 114, 140, 143, 165, 194, 208
Resources recovery facility, 78–80, 145, 152, 159, 161, 168, 170, 175, 176, 180, 181, 190, 197, 198, 200
Resources Recovery (Dade County) Inc., 78–80
Richmond (California), 85–7
Roads, 26, 59, 115, 162, 188
Royal Commission on Environmental Pollution (RCEP), 5, 22, 95, 105, 106, 120, 121, 192, 218
Royal Town Planning Institute, 112

Salford, 131
San Diego, 53
San Francisco, 52, 83, 87, 88, 150, 151, 165, 199
Sanctions, 161, 191, 199

Scandinavia, 93
Scoping, 64, 65, 215
Scotland, 93, 94
Scrubber, 25, 83, 210
Secretary of State for the Environment, 100, 104, 110, 111, 113, 114, 127, 133, 134, 136, 138
Sewage sludge incinerator, 126–8, 145, 154, 155, 159, 161, 163, 165, 178, 179, 186, 188, 190, 193, 195, 208
Shellstar Ltd, 133–5
Sierra Club, 37, 88, 161, 201
Simplified planning zones, 108, 112, 123, 196
Siting process, 2, 3–5, 10, 14, 28, 39, 140, 141, 143, 148, 155, 215
Skyview Addition, 74–6
Smoke, 56, 95, 97, 98, 103, 111, 119, 121, 168
SOHIO pipeline, 38
Solano County (California), 87–9
Solvent recycling plant, 81–3, 144, 145, 147, 156, 158, 159, 162, 163, 164, 165, 169, 170, 171, 175, 176, 178, 180, 186, 187, 190, 192, 195, 198, 199, 200, 201, 210
Spectron Inc., 81–3
St Helens, 135–7
St Tammary Parish (Louisiana), 71–4
Standard Zoning Enabling Act 1924 (US), 52
Standards, 21–3, 42, 43, 45, 46, 47, 55, 63, 69, 79, 88, 107, 111, 118, 132, 148, 151, 196, 211, 218, 219, 220, 222
State:
  implementation plans, 42, 43, 44, 46
  land use controls, 53, 57–61
Sterling Mouldings Ltd, 125–6
Stewardship ethic, 91, 92
Subdivision control, 55, 56, 57
Sulphur dioxide, 9, 19, 20, 25, 41, 42, 43, 45, 46, 47, 49, 79, 84, 88, 97, 98, 103, 119, 121, 135, 137, 138, 139, 147, 168, 177, 195, 210
Sulphuric acid works, 135–7, 147, 150, 156, 161, 163, 168, 169, 170, 171, 175, 176, 182, 183, 191, 197, 199, 200, 201, 206

Sustainable development, 2
Sweden, 209

Tameside, 125–6
Tax, 35, 37, 57, 70, 73, 78, 83, 84, 85, 112, 153, 176, 187, 188, 212, 217, 219
Technical controls, 2, 14, 23, 25–6, 27, 31, 99
Texas, 38, 54, 61, 70, 74–6
Topography, 28, 158, 214
Town and Country Planning Acts (UK), 108, 111, 114
Trades Union Congress, 100
Transaction costs, 38, 95, 148, 204, 205, 206, 211
Transport, 16, 45, 54, 69
Transport and General Workers Union, 134, 146

UKF Ltd, 133–5
Unwin, 108
Use Classes Order (UK), 111, 129, 221
US Steel, 167
Utilitarianism, 34, 92

Vale Royal, 133

Vegetation, 27, 29, 30, 43, 46, 93, 127, 133, 159
Vermont, 59
Vigilance of public, 171, 201
Virginia, 59
Visibility, 20, 42, 46
Volatile organic components, 41, 46, 49

Wales, 93, 94
Washington, 77
Waste, 9, 25, 26, 27, 77, 110, 115
Water pollution, 74, 77, 167
West Linn (Oregon), 85
West Yorkshire, 128–30
Wheelabrator Frye Inc., 83–5
Wilderness, 41, 45
World Health Organisation (WHO), 22
Worsley, 131
Wrightsville Beach (North Carolina), 78

Zoning, 23, 52, 55–6, 60, 61, 62, 74, 75, 82, 108, 123, 152, 155, 178, 185, 186, 187, 192, 196, 208, 217
Ordinances, 34, 53, 55, 61, 70, 81, 85, 189